AIRCRAFT INSTRUMENTS

AIRCRAFT INSTRUMENTS

Principles and Applications

E H J Pallett
AFSLAET, T Eng (CEI), ARAeS, MIN

with a foreword by
Air Cdre Sir Vernon Brown

Pitman

PITMAN PUBLISHING LIMITED
39 Parker Street, London WC2B 5PB

Associated Companies
Copp Clark Pitman, Toronto
Fearon Pitman Publishers Inc, San Francisco
Pitman Publishing New Zealand Ltd, Wellington
Pitman Publishing Pty Ltd, Melbourne

© E H J Pallett 1972

First published in Great Britain 1972
Reprinted 1977, 1978, 1979

All rights reserved. No part of this publication may be reproduced,
stored in a retrieval system, or transmitted, in any form or by any
means, electronic, mechanical, photocopying, recording and/or
otherwise without the prior written permission of the publishers.
This book may not be lent, resold, hired out or otherwise disposed of
by way of trade in any form of binding or cover other than that in
which it is published, without the prior consent of the publishers.
This book is sold subject to the Standard Conditions of Sale
of Net Books and may not be resold in the UK below the net price.

Printed and Bound in Great Britain by
Spottiswoode Ballantyne Ltd,
Colchester and London

ISBN 0 273 31747 4

Foreword

by Air Cdre Sir Vernon Brown CB, OBE, MA,
CEng, FRAeS, HonFSLAET

The aeroplane on which I took my Aero Club ticket had as its instrumentation an engine rev. counter and an oil pressure gauge, an altimeter which recorded up to 16,000 ft (and mighty slow it was) and an airspeed indicator on which the scale showed up to about 75 m.p.h. (the top speed of a Maurice Farman Longhorn was plus/minus 60 m.p.h. and its best climbing speed about 28 m.p.h.). This airspeed indicator consisted of a small cup on the end of a rod at the end of which was a bell crank lever which actuated a pointer. As it was mounted on the outer strut, to see one's speed at all from the front tandem seat meant turning one's head more than 90° to the left!

A lot of water has flowed under the world's bridges since early 1915, and with the advent of the cabin type of aircraft and the turbine engine the aeroplane has become a very complicated vessel indeed, the main purposes of which are either to carry enormous loads of highly flammable fuel at great speed and height with as big a payload as possible, or simply to destroy an enemy before he destroys you. Naturally its instrumentation has become proportionally difficult to install and maintain.

This book by Mr Pallett has been written with the aim of helping those whose job is to keep aircraft in the air to understand their instruments and, of course, the fundamental principles underlying their design and application to flight.

Every time I am invited to an aircraft pilot's cabin, I thank my stars I no longer fly except as a passenger and my sympathy goes out to the licenced aircraft engineers who are responsible for ensuring that everything works according to plan. Theirs is a very difficult job, and if Mr Pallett's book will help them, as indeed I feel sure it must, then he will have done a good job and will be the recipient of very many blessings.

Preface

The steady growth in the number and scope of aircraft instruments has run parallel with the complex growth of aircraft themselves, and in the development of methods of detecting, processing and presenting relevant control information, the design and constructional patterns of instruments have likewise grown in a complex fashion. As a result, instruments are often associated with a science veiled in considerable "electrickery" contained within numerous black boxes produced by obscure persons posing as disciples of Pandora!

There is, of course, no denying the abundance of black boxes (they are invariably grey these days anyway!) and the many functional and constructional changes, but as in the evolutionary processes of most technological fields, complexities arise more often than not in developing new methods of applying old but well-established principles. For example, in the measurement of altitude, airspeed and turbine engine thrust, the appropriate pressures are detected and measured by the deflections of capsule- and diaphragm-type detecting elements just as they were in some of the first instruments ever developed for pressure measurement in aircraft. Similarly, instruments employed for the measurement and control of engine temperatures still depend for their operation on the changes in electrical resistance and thermo-electric characteristics of certain metals under varying temperature conditions.

Thus, in preparing the material for this book, emphasis has been placed on fundamental principles and their applications to flight, navigation and engine-performance monitoring instruments. It has not been possible to include every type of instrument, but it is considered that the coverage is representative of a wide range of current aircraft instrument installations, and should provide a firm foundation on which to base further study.

The material is arranged in a sequence which the author has found useful in the implementation of training programmes based on relevant sections of the various examination syllabuses established for aircraft maintenance engineers, and the various ratings of pilots' licences. In this connection it is therefore hoped that the book will prove a useful source of reference for the experienced instructor as well as for the student. A selection of questions are given at the end of each chapter and the author is indebted to the Society of Licensed Aircraft Engineers and Technologists for permission to reproduce questions selected from examination papers.

Valuable assistance has been given by a number of organizations in supplying technical data, and in granting permission to reproduce many of the illustrations. Acknowledgement is hereby made to the following: Smith's Industries Ltd, Aviation Division; Sperry Rand Ltd, Sperry Gyroscope Division; Sangamo-Weston Ltd; Thorn Bendix; British Overseas Airways Corporation; British Aircraft Corporation (Operating) Ltd; R. W. Munro Ltd; Dowty Electrics Ltd; Negretti and Zambra (Aviation) Ltd; Hawker-Siddeley Aviation Ltd.

Copthorne   E.P.
Sussex

Contents

	Foreword	v
	Preface	vii
	Historical Background	1
1	Requirements and Standards	7
2	Instrument Elements and Mechanisms	11
3	Instrument Displays, Panels and Layouts	23
4	Pitot-static Instruments and Systems	55
5	Primary Flight Instruments (Attitude Indication)	111
6	Primary Heading Indicating Instruments	150
7	Remote-indicating Compasses	178
8	Aircraft Magnetism and its Effects on Compasses	203
9	Accelerometers and Fatigue Meters	223
10	Synchronous Data-transmission Systems	230
11	Measurement of Engine Speed	252
12	Measurement of Temperature	265
13	Measurement of Pressure	296
14	Measurement of Fuel Quantity and Fuel Flow	312
15	Engine Power and Control Instruments	335
16	Integrated Instrument and Flight Director Systems	354
	List of Symbols	362
	Solutions to Numerical Questions	363
	Index	365

Historical Background

In the days of the first successful aeroplanes the problems of operating them and their engines according to strict and complicated procedures, of navigating over long distances day or night under all weather conditions, were, of course, problems of the future. They were, no doubt, envisaged by the then enthusiastic pioneers of flight, but were perhaps somewhat overshadowed by the thrills of taking to the air, manoeuvring and landing.

Such aeroplanes as these pioneers flew were rather "stick and string" affairs with somewhat temperamental engines, the whole combination being manoeuvered by a pilot lying, sitting or crouching precariously in the open, for the luxury of a cockpit was also still to come. Instruments designed specifically for use in an aeroplane were also non-existent; after all, what instrument manufacturer at the time had had the necessity of designing, for example, an instrument to show how fast a man and a machine could travel through the air?

It is a little difficult to say exactly in what sequence instruments were introduced into aeroplanes. A magnetic compass was certainly an early acquisition as soon as pilots attempted to fly from A to B, and flying greater distances would have required information as to how much petrol was in the tank, so a contents gauge was fitted usually taking the form of a glass sight gauge. Somewhere along the line the clock found its place and was useful as a means of calculating speed from a time/distance method, and as an aid to navigation. With such supplementary aids a pilot was able to go off into the third dimension flying mainly by his direct senses and afterwards boasting perhaps that instruments would not be needed anyway!

As other pioneers entered the field many diverse aeroplane designs appeared some of which were provided with an enclosure for the pilot and a wooden board on which the then available instruments could be mounted. Thus the cockpit and instrument panel were born.

Shortly before the outbreak of World War I, some attention was given to the development of instruments for use on military and naval aeroplanes, and the first principles of air navigation were emerging with designs for instruments specially adapted for the purpose. Consequently, a few more instruments appeared on the dashboards of certain types of aeroplane including an altimeter, airspeed indicator and the first engine instruments—an r.p.m. indicator and an oil pressure gauge.

During the war years very few new instruments were provided in the many types of aeroplane produced. A requirement did arise for the indication of an aeroplane's pitch and bank attitude, which led to the introduction of the fore-and-aft level and the cross-level. The former instrument consisted of a specially constructed glass tube containing a liquid which moved up and down against a graduated scale, and the latter was a specially adapted version of the simple spirit level.

The main progress of the war years as far as instruments were concerned was in the development of the existing types to higher standards of accuracy, investigation of new principles, and the realization that instruments had to be designed specifically to withstand vibration, acceleration, temperature change, and so on. It was during this period that aircraft instruments became a separate but definite branch of aviation.

After the war, aviation entered what may be termed its second pioneering stage in which ex-wartime pilots flew air routes never before attempted. In 1919, for example, Alcock and Brown made the first non-stop Atlantic crossing; in the same year the Australian brothers Keith and Ross Smith made the first flight from England to Australia. Flights such as these and others carried out in the '20s, were made in military aircraft and with the aid of the same instrument types as had been used in wartime. Although these flights laid the foundation for the commercial operation of the aeroplane, it was soon realized that this could not be fully exploited until flights could be safely carried out day and night and under adverse weather conditions. It had already been found that pilots soon lost their sense of equilibrium and had difficulty in controlling an aeroplane when external references were obscured. Instruments were therefore required to assist the pilot in circumstances which became known as "blind flying conditions."

The first and most important step in this direction was the development of the turn indicator based on the principles of the gyroscope. This instrument, in conjunction with the magnetic compass, became an extremely useful blind-flying aid, and when a bank indicator was later added to the turn indicator, pilots were able, with much patience and skill, to fly "blind" by means of a small group of instruments.

However, progress in the design of aeroplanes and engines developed to a stage where it was essential to provide more aids to further the art of blind flight. An instrument was required which could replace the natural horizon reference and could integrate the information hitherto obtained from the cross-level and the fore-and-aft level. It was also necessary to have some stable indication of heading which would not be affected by acceleration and turning —manoeuvres which had for long been a source of serious errors in the magnetic compass.

The outcome of investigations into the problem was the introduction of two more instruments utilizing gyroscopic principles, namely the gyro horizon and the directional gyro, both of which were successfully proved in the first-ever instrument flight in 1929. At this time the sensitive altimeter and the rate-of-climb indicator had also appeared on the instrument panel ("dashboard" was now rather a crude term!), together with more engine instruments. Engines were being supercharged and so the "boost" pressure gauge came into vogue; temperatures of oil and liquid cooling systems and fuel pressures were required to be known, and consequently another problem arose—instrument panels were getting a little overcrowded. Furthermore, these instruments, essential though they were, being grouped on the panel in a rather haphazard manner, and this made it somewhat difficult for pilots to assimilate the indications, to interpret them and to base on them a definite course of action.

Thus, by about the middle '30s grouping of instruments became more rationalized so that "scanning distance" between instruments was reduced to a minimum. The most notable result of rationalization was the introduction of the separate "blind flying panel" containing the airspeed indicator, altimeter, gyro horizon, directional gyro, rate-of-climb indicator (vertical speed) and turn-and-bank indicator. This method of grouping the flight instruments has continued up to the present day.

Continuing developments in military and commercial aviation brought about faster and bigger aeroplanes, multi-engine arrangements, retractable landing gear systems, electrical systems, etc., all of which called for more and more instruments. The instrument designer kept pace with these developments by introducing electrically operated instruments and systems of remote indication, but the fitting of knobs, switches and handles for the operation of other systems imposed restrictions on space. Pilots were therefore once again facing problems of assimilating the indications of haphazardly placed instruments. In multi-engined aircraft the problems were further aggravated.

The greater ranges of multi-engined aircraft meant longer periods in the air and presented the problem of pilot fatigue: a problem not unknown to the pioneers of long-distance flying. This was alleviated by equipping bomber and long-range commercial aircraft of the late '30s, with an automatic pilot, a device which had been successfully demonstrated as far back as 1917. With the automatic pilot in operation, pilots were enabled to devote more attention to instrument monitoring, and to the involved navigational and radio communication techniques which had also been introduced.

A further step in relieving the pilot's work load was made when navigator, radio operator and flight engineer stations were introduced, becoming standard features during World War II. It thus became possible to mount the instruments appropriate to the crew member's duties on separate panels at his station, leaving the pilot with the instruments essential for the flight handling of the aircraft.

One of the most outstanding developments resulting from the war years was in the field of navigation, giving rise notably to the full-scale use of the remote-transmitting compass system in conjunction with such instruments as air and ground position indicators, and air mileage units. Flight instruments had been improved, most instruments for engine operation were now designed for electrical operation, and as an aid to conserving panel space more dual-type instruments had been introduced.

Another development which took place, and one which changed the picture of aviation, was that of the gas turbine engine. As a prime mover, it opened up many possibilities: more power could be made available, greater speeds and altitudes were possible, aircraft could be made "cleaner" aerodynamically; and being simpler in its operation than the piston engine, the systems required for its operation could also be made simpler. From the instrument point of view the changeover was gradual and initially did not create a sudden demand for completely new types of instrument. The rotational speeds of turbine engines were much higher than those of piston engines and so r.p.m. indicators had to be changed accordingly, and a new parameter, gas temperature, came into existense which necessitated an additional thermometer, but apart from these two, existing engine and flight instruments could still be utilized.

This state of affairs, however, lasted for only a few years after the war. As the aircraft industry geared itself up into its peacetime role, various designs of gas-turbine powered aircraft went into production and flew alongside the more conventional types. Demands for greater speeds and altitudes meant bigger

Blackburn monoplane (1910)

BOAC 747 (1971)

engines, and as new high-temperature materials and newer systems were developed, the required power was produced, but the "jet" was losing a lot of its simplicity in the process!

Once again instruments had to be modified and new designs introduced. For example, the turbine engine produced far less vibration at the instrument panel, and as a result there was a tendency for the slight inherent static friction of an instrument mechanism to "stick" the pointers. The mechanisms were therefore designed to make them function correctly and continuously in the absence of vibration. Increases of speed brought about an effect known as compressibility, and to warn pilots of their approach to a dangerous flight situation, the speed indicator known as the Machmeter was introduced.

The higher altitudes made possible by the turbine engine, and the fast controlled descent procedures later adopted, necessitated the use of altimeters having extended ranges, and scales which could be read easily and without ambiguity. In order to meet these requirements and to help pilots avoid subsequent "mis-reading incidents," the altimeter has passed, and continues to pass, through various stages of modification.

In the aircraft electrical and electronic field advances were also being made so that new measuring techniques were possible for instruments; for example, the measurement of fuel quantity by means of special types of capacitor located in the fuel tanks, electrically-operated gyro horizons and turn and bank indicators providing for greater stability and better performance at high altitudes than their air-driven counterparts.

As a result of rapid growth of radio aids to navigation, specific "radio aid" instruments were also introduced to present additional information for use in conjunction with that provided by the standard flight and navigation instruments. Although essential for the safe operation of aircraft, particularly during the approach and landing phases of flight, the pilot's workload was increased and it was foreseen that eventually the locating of separate instruments on panels would once more become a problem. It was therefore natural for an integration technique to be developed whereby the data from a number of instrument sources could be presented in a single display. Thus, integrated flight instrument and flight director systems were evolved, and are now a standard instrumentation feature of many types of aircraft currently in service, not only for the display of primary flight data, but also for the monitoring of advanced automatic flight control systems.

The display of more varied data covering the performance of engines and systems also became necessary with the growing complexity of engines, and so the number of separate indicators increased. However, the application of minaturized electrical and electronic components, and micro-circuit techniques has permitted large reductions in the dimensions of instrument cases thereby helping to keep panel space requirements within reasonable bounds. Furthermore, it has led to the up-dating of an early data presentation method, namely the vertical scale, for the engine instruments of a number of today's aircraft. Another technique now applied as a standard feature of turbine engine instrumentation is based on one successfully developed in the late '40s, i.e. the control of engine speed and gas temperature by the automatic regulation of fuel flow. In this technique the signals generated by the standard tachometer generators and by thermocouples, are also processed electronically and are used to position the appropriate fuel control valve system.

From this very brief outline of instrument development it will be particularly noted that this has for the most part been based on "a quart into a pint pot" philosophy, and has been continuously directed to improving the methods of presenting relevent data. This has been a natural progression, and development in these areas will continue to be of the utmost importance, paralleling as it does the increasing complexity of aircraft and their systems. As the controller of a "man-machine loop" and in developing and operating the machine, man has been continually reminded of the limitations of his natural means of sensing and processing control information. However, in the scientific and technical evolutionary processes, instrument layout design and data presentation methods have become a specialized part of ergonomics, or the study of man in his working environment; this together with the rapid strides made in avionics, culminates in the provision of electronic display instruments, computerized measuring elements, integrated instrument and flight control systems for fully automatic control, enabling man to deal with an expanding task within the normal range of human performance.

1 Requirements and Standards

The complexity of modern aircraft and all allied equipment, and the nature of the environmental conditions under which they must operate, requires conformity of design, development and subsequent operation with established requirements and standards. This is, of course, in keeping with other branches of mechanical and transport engineering, but in aviation requirements and standards are unique and by far the more stringent.

The formulation and control of airworthiness requirements as they are called, and the recommended standards to which raw materials, instruments and other equipment should be designed and manufactured, are established in the countries of design origin, manufacture and registration, by government departments and/or other legally constituted bodies. The international operation of civil aircraft necessitates international recognition that aircraft do, in fact, comply with their respective national airworthiness requirements. As a result, international standards of airworthiness are also laid down by the International Civil Aviation Organization (ICAO). These standards do not replace national regulations, but serve to define the complete minimum international basis for the recognition by countries of airworthiness certification.

It is not the intention that we should go into all the requirements—these take up volumes in themselves—but rather to extract some essential requirements related to instruments; by so doing we can provide ourselves with a very useful foundation on which to base a study of the operating principles of instruments and how they are applied in meeting the requirements.

REQUIREMENTS

Location, Visibility and Grouping of Instruments

1. All instruments shall be located so that they can be read easily by the appropriate member of the flight crew.
2. When illumination of instruments is provided there shall be sufficient illumination to make them easily readable and discernible by night. Instrument lights shall be installed in such a manner that the pilot's eyes are shielded from their direct rays and that no objectionable reflections are visible to him.
3. Flight, navigation and power-plant instruments for use by a pilot shall be plainly visible to him from his station with the minimum practicable deviation from his normal position and line of vision when he is looking out and forward along the flight path of the aircraft.
4. All flight instruments shall be grouped on the instrument panel and, as far as practicable, symmetrically disposed about the vertical plane of the pilot's forward vision.
5. All the required power-plant instruments shall be conveniently grouped on instrument panels and in such a manner that they may be readily seen by the appropriate crew member.

6. In multi-engined aircraft, identical power-plant instruments for the several engines shall be located so as to prevent any misleading impression as to the engines to which they relate.

Instrument Panels

The vibration characteristics of instrument panels shall be such as not to impair seriously the accuracy of the instruments or to damage them. The minimum acceptable vibration insulation characteristics are established by standards formulated by the appropriate organization; for example in the United Kingdom this means the British Standards Institution, the data being contained in BS G.141, "Flexibly mounted instrument panels for fixed wing aircraft."

Instruments to be Installed

Flight and Navigation Instruments

1. Altimeter adjustable for changes in barometric pressure
2. Airspeed indicator
3. Vertical speed indicator
4. Gyroscopic bank-and-pitch attitude indicator
5. Gyroscopic rate-of-turn indicator (with bank indicator)
6. Gyroscopic direction indicator
7. Magnetic compass
8. Outside air temperature indicator
9. Clock

Pitot-static System

Instruments 1, 2 and 3 form part of an aircraft pitot-static system, which must also conform to certain requirements. These are summarized as follows:

(a) The system shall be air-tight, except for the vents to atmosphere, and shall be arranged so that the accuracy of the instruments cannot be seriously affected by the aeroplane's speed, attitude, or configuration; by moisture, or other foreign matter.
(b) The system shall be provided with a heated pitot-tube or equivalent means of preventing malfunctioning due to icing.
(c) Sufficient moisture traps shall be installed to ensure positive drainage throughout the whole of the system.
(d) In aircraft in which an alternate or emergency system is to be installed, the system must be as reliable as the primary one and any selector valve must be clearly marked to indicate which system is in use.
(e) Pipelines shall be of such an internal diameter that pressure lag and possibility of moisture blockage is kept to an acceptable minimum.
(f) Where static vents are used, to obviate yawing errors they shall be situated on opposite sides of the aeroplane and connected together as one system. Where duplicate systems are prescribed, a second similar system shall be provided.

Gyroscopic Instruments

Gyroscopic instruments may be of the vacuum-operated or electrically operated type, but in all cases the instruments shall be provided with two independent sources of power, a means of selecting either power source, and a means of indicating that the power supply is working satisfactorily.

The installation and power supply system shall be such that failure of one instrument, or of the supply from one source, or a fault in any part of the supply system, will not interfere with the proper supply of power from the other source.

Duplicate Instruments

In aircraft involving two-pilot operation it is necessary for each pilot to have his own pitot-static and gyroscopic instruments. Therefore two independent operating systems must be provided and must be so arranged that no fault, which might impair the operation of one, is likely to impair the operation of both.

Magnetic Compass

The magnetic compass shall be installed so that its accuracy will not be excessively affected by the aeroplane's vibration or magnetic fields of a permanent or transient nature.

Power Plant Instruments

1. Tachometer to measure the rotational speed of a crankshaft or a compressor as appropriate to the type of power plant.
2. Coolant temperature indicator for a liquid-cooled engine.
3. Cylinder-head temperature indicator for an air-cooled engine to indicate the temperature of the hottest cylinder.
4. Carburettor-intake air temperature indicator.
5. Oil temperature indicator to show the oil inlet and/or outlet temperature.
6. For turbojet and turbopropeller engines a jet-pipe temperature indicator or equivalent to indicate whether the exhaust gas temperature is maintained within its limitations.
7. Fuel-pressure indicator to indicate pressure under which fuel is being supplied and a means for warning of low pressure.
8. Oil-pressure indicator to indicate pressure under which oil is being supplied to a lubricating system and a means for warning of low pressure.
9. Manifold pressure gauge for a supercharged engine.
10. Fuel-quantity indicator to indicate in gallons or equivalent units the quantity of usable fuel in each tank during flight. Indicators shall be calibrated to read zero during cruising level flight, when the quantity of fuel remaining is equal to the unusable fuel, i.e. the amount of fuel remaining when, under the most adverse conditions, the first evidence of malfunctioning of an engine occurs.
11. Fuel-flow indicator for turbojet and turbopropeller engines. For piston engines, not equipped with an automatic mixture control, a fuel flowmeter or fuel/air ratio indicator.
12. Thrust indicator for a turbojet engine.
13. Torque indicator for a turbopropeller engine.

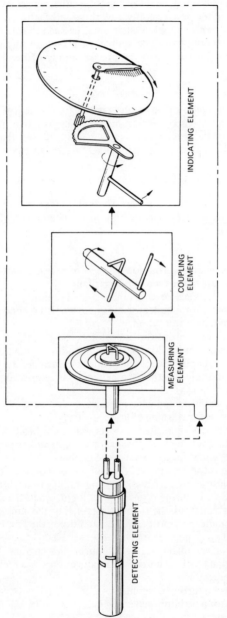

Fig 2.1 Elements of an instrument

2 Instrument Elements and Mechanisms

ELEMENTS

From the operating point of view, we may regard an instrument as being made up of the following four principal elements: (i) the *detecting element*, which detects changes in value of the physical quantity or condition presented to it; (ii) the *measuring element*, which actually measures the value of the physical quantity or condition in terms of small translational or angular displacements; (iii) the *coupling element*, by which displacements are magnified and transmitted; and (iv) the *indicating element*, which exhibits the value of the measured quantity transmitted by the coupling element, by the relative positions of a pointer, or index, and a scale. The relationship between the four elements is shown in Fig 2.1.

MECHANISMS

In the strictest sense, the term *mechanism* refers to all four elements as a composite unit and contained within the case of an instrument. However, since the manner in which the functions of the elements are performed and integrated is governed by relevant instrument operating principles and construction, this applies to only a very few instruments. In the majority of applications to aircraft, a separation of some of the elements is necessary so that three, or maybe only two, elements form the mechanism within the instrument case. The direct-reading pressure gauge shown at (*a*) in Fig 2.2 is a good example of a composite unit of mechanical elements, while an example of separated mechanical elements as applied to an airspeed indicator is shown at (*b*). In this example the detecting element is separated from the three other elements, which thus form the mechanism within the case.

There are other examples which will become evident as we study subsequent chapters, but at this stage it will not be out of place to consider the operation of a class of mechanisms based on the principles of levers and rods. These are utilized as coupling elements which follow definite laws, and can introduce any required input/output relationship. In aircraft instrument applications, such lever and rod mechanisms are confined principally to direct-reading pressure gauges and pitot-static flight instruments.

Lever Mechanism

Let us consider first of all the simple Bourdon tube pressure gauge shown at (*a*) in Fig 2.3. The Bourdon tube forms both the detecting and measuring elements, a simple link, lever, quadrant and pinion forms the coupling element, while the indicating element is made up of the pointer and scale. This mechanism is of the basic lever type, the lever being in this case the complete coupling element. When pressure is applied to the tube it is displaced, such

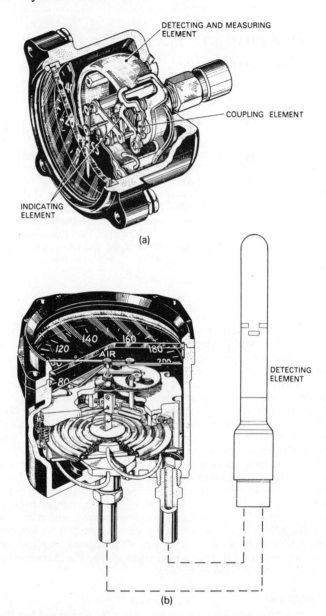

Fig 2.2 Instrument mechanisms
(a) Direct-reading pressure gauge
(b) Airspeed indicator containing measuring, coupling and indicating elements

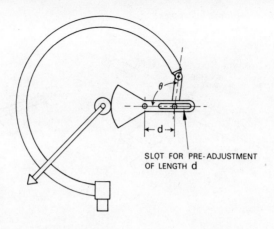

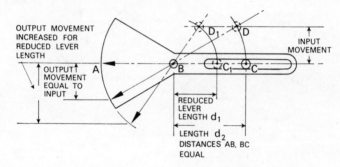

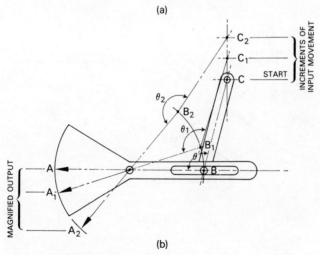

Fig 2.3 Simple lever mechanism
(a) Effect of lever length
(b) Effect of lever angle on magnification

14 *Aircraft Instrument*

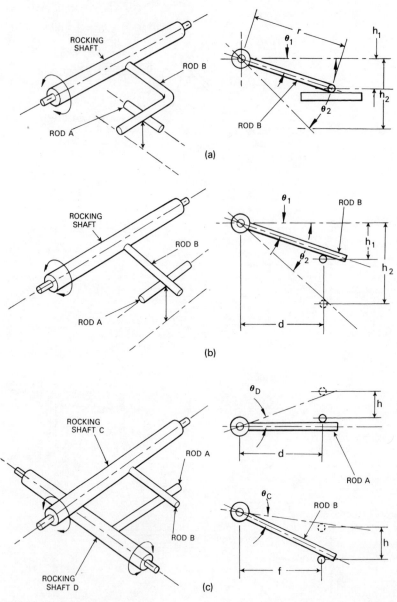

Fig 2.4 Rod mechanism

(a) Sine mechanism (b) Tangent mechanism (c) Double tangent mechanism

displacement resulting in input and output movements of the coupling and indicating elements, respectively, in the directions shown.

In connection with mechanisms of this type, two terms are used both of which are related to the movement and calibration of the indicating element; they are, *lever length*, which is the distance d between the point of operation of the measuring element and the pivoting point of the lever, and *lever angle*, which is the angle θ between the lever and the link connecting it to the measuring element.

In order to understand what effects these have on the input/output relationship, let us again refer to Fig 2.3 (*a*). The movement of the indicating element is proportional to the lever length; thus, if the lever is pivoted at its centre, this movement will be equal to the input movement. Let us now assume that the pivoting point is moved to a distance d_1 from the point of operation. The lever length is now reduced so that for the same input movement as before the output movement of the indicating element will be increased. From this it will be clear that an increase of lever length to a distance d_2 will produce a decreased output movement for the same input movement.

The effect of lever angle on the input/output relationship is to change the rate of magnification since the lever angle itself changes in response to displacement of the measuring element. This effect is evident from Fig 2.3 (*b*). If we assume that the line AB represents the axis of the lever at its starting position, then the starting lever angle will be θ. Assume now that the measuring element is being displaced by equal increments of pressure applied to it. The link attachment point C will move to C_1 and will increase the lever angle in two stages; firstly when the link pivots about point B, and secondly when the link pulls the lever arm of the coupling element upwards from the starting position taking point B to point B_1. Thus, the axis of the lever arm has moved to A_1B_1 and the lever angle has increased to a total angle θ_1. When the next increment of pressure is applied, point C reaches C_2, point B reaches B_2 and the axis AB moves to A_2B_2, so that, not only has the lever angle been further increased, but also the magnification, the distance from A_1 to A_2 being much greater than that from A to A_1.

From the foregoing, it would appear that the two effects counteract each other, and that erratic indications would result. In all instruments employing lever mechanisms, however, provision is made for the adjustment of lever lengths and angles so that the indicating element follows the required calibration law within the limits permissible.

Rod Mechanisms

Unlike pure lever mechanisms, rod mechanisms dispense with pin or screw-jointed linkages for the interconnection of component parts, and rely on rods in contact with, and sliding relative to, each other for the generation of the input/output relationship. Contact between the rods under all operating conditions is maintained by the use of a hairspring which tensions the whole mechanism.

These mechanisms, shown in Fig 2.4, find their greatest application in flight instruments, and can be divided into three main classes named after the trigonometrical relationships governing their operation. They are: (i) the sine mechanism, (ii) the tangent mechanism and (iii) the double-tangent mechanism.

The *sine mechanism*, Fig 2.4 (a), is employed in certain types of airspeed indicator as the first stage of the coupling element, and comprises two rods A and B in sliding contact with each other, and a rocking shaft C to which rod B is attached. In response to displacement of the measuring element, the input movement of rod A is in a vertical plane, causing rod B to slide along it and at the same time to rotate the rocking shaft. The point of contact between the two rods remains at a constant radius r from the centre of the rocking shaft.

The rotation of the rocking shaft is given by the trigonometrical relationship

$$h_2 - h_1 = r(\sin\theta_2 - \sin\theta_1)$$

where h is the vertical input movement of rod A and θ the angle of rod B. The usable range of movement (θ in the diagram) of rod B is $\pm 60°$, and the angle at which it starts within this range depends on the magnification required for calibration. For example, if rod A moves upwards from a starting angle at $-60°$, the magnification is at first high and then decreases with continued movement of rod A. When the starting angle is at or near the zero degree position, the magnification rate is an increasing one.

A *tangent mechanism* is similar to a sine mechanism, but as will be noted from Fig 2.4 (b), the point of contact between the two rods remains at a constant perpendicular distance d from the centre of the rocking shaft. The rotation of the rocking shaft is given by the relationship

$$h_2 - h_1 = d(\tan\theta_2 - \tan\theta_1)$$

The magnification rate of this mechanism is opposite to that of a sine mechanism except at a starting angle at or near zero, where $\sin\theta$ and $\tan\theta$ are approximately equal.

Figure 2.4 (c) illustrates a *double-tangent mechanism*, which is employed where rotary motion of a shaft is to be transferred through a right angle. A typical application is as the second stage of an airspeed indicator coupling element and for the gearing of the indicating element.

As will be evident from the diagram, it is formed basically of two tangent mechanisms in series so that the rotary motion of one shaft is converted into rotary motion of a second. This is instead of converting a linear motion into a rotary one as with a sine or a tangent mechanism. The input/output relationship is a combined one involving two trigonometrical conversions; the first is related to the movement of the contact point between rods A and B and is given by

$$h = d(\tan\theta_{D2} - \tan\theta_{D1})$$

where d is the perpendicular distance between the axis of shaft D and the plane of contact between A and B, and θ_D is the rotation of D. The second conversion is given by

$$h = f(\tan\theta_{C2} - \tan\theta_{C1})$$

where f is the perpendicular distance between the axis of shaft C and the plane of contact between A and B, and θ_C is the rotation of shaft C. When the planes of movement of rods A and B intersect at right angles and the rods are straight ones, the combination of the two conversions gives the relationship:

$$d(\tan\theta_{D1} - \tan\theta_{D2}) = f(\tan\theta_{C1} - \tan\theta_{C2}).$$

A variation on the double tangent theme, is the *skew tangent mechanism*. In this, the rocking shafts are orthogonal but the planes of the rods A and B do not intersect at right angles.

GEARS

The coupling and indicating elements of many aircraft instruments employ gears in one form or another, for the direct conversion of straight-line or arc-like motion into full rotary motion, and for increasing or decreasing the motion. Figure 2.5 illustrates in schematic form how gears are applied to an

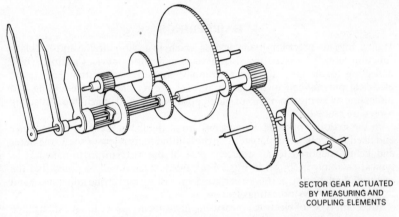

SECTOR GEAR ACTUATED BY MEASURING AND COUPLING ELEMENTS

Fig 2.5 Gear assembly for a multi-pointer indicating element

instrument utilizing a multi-pointer type of indicating element. The sector gear and its meshing pinion provide for the initial magnification of the measuring element's displacement. The gear is a small portion of a large geared wheel, and since it has as many teeth in a few degrees of arc as the pinion has completely around it, the sector need only turn a few degrees to rotate the pinion through a complete revolution. The other gears shown in Fig 2.5 are designed to provide a definite magnification ratio of movement between their respective pointers and the pointer actuated by the sector gear and pinion.

In applying gears to instruments and control systems, a problem which has to be faced is that a gear can always turn a small amount before it will drive the one in mesh with it. This loss of motion, or *backlash* as it is termed, is unavoidable since the dimensioning of the gear teeth must allow for a set amount of "play" to avoid jamming of the gears. Other methods must therefore be found to minimize the unstable effects which backlash can create.

The method most commonly adopted in geared mechanisms is one involving the use of a coiled hairspring. The hairspring usually forms part of an indicating element and is positioned so that one end is attached to the pointer shaft and the other to the mechanism frame. In operation, the spring due to tensioning always has a tendency to unwind so that the inherent play between gear teeth is taken up and they are maintained in contact.

Another method, and one which is adopted in certain instrument systems involving the transmission of data, is the *anti-backlash gear*. This consists of two identical gears freely mounted face to face on a common hub and interconnected with each other by means of two springs so that, in effect, it is a split single gear wheel. Before the gear is meshed with its partner, one half is rotated one or two teeth thus slightly stretching the springs. After meshing, the springs always tend to return the two halves of the gear to the static unloaded position; therefore the faces of all teeth are maintained in contact. The torque exerted by the springs is always greater than the operating torques of the transmission system so that resilience necessary for gear action is unaffected.

HAIRSPRINGS

Hairsprings are precision-made devices which, in addition to the anti-backlash function already referred to, also serve as controlling devices against which deflecting forces are balanced to establish required calibration laws (as in electrical moving-coil instruments) and for the restoration of coupling and indicating elements to their original positions as and when the deflecting forces are removed.

In the majority of cases, hairsprings are of the flat-coil type with the inner end fixed to a collet, enabling it to be press-fitted to its relevant shaft, the outer end being anchored to an adjacent part of the mechanism framework. A typical assembly is shown in Fig 2.6 (*a*), from which it will be noted that the method of anchoring permits a certain degree of spring torque adjustment and initial setting of the indicating element.

In certain types of electrical measuring instruments, provision must be made for external adjustment of the pointer to the zero position of the scale. One method commonly adopted, and which illustrates the principles in general, is shown in Fig 2.6 (*b*). The inner end of the spring is secured to the pointer shaft in the normal way, but the outer end is secured to a circular plate friction-loaded around the front pivot screw. A fork, which is an integral part of the plate, engages with a pin eccentrically mounted in a screw at the front of the instrument. When the screw is rotated it deflects the plate thus rotating the spring, shaft and pointer to a new position without altering the torque loading of the spring.

The materials from which hairsprings are made are generally phosphor-bronze and beryllium-copper, their manufacture calling for accurate control and grading of thickness, diameter and torque loading to suit the operating characteristics of particular classes of instrument.

TEMPERATURE COMPENSATION OF INSTRUMENT MECHANISMS

In the construction of instrument mechanisms, various metals and alloys are used, and unavoidably, changes in their physical characteristics can occur with changes in the temperature of their surroundings. For some applications deliberate advantage can be taken of these changes as the basis of operation; for example, in certain electrical thermometers the changes in a metal's electrical resistance forms the basis of temperature measurement. However,

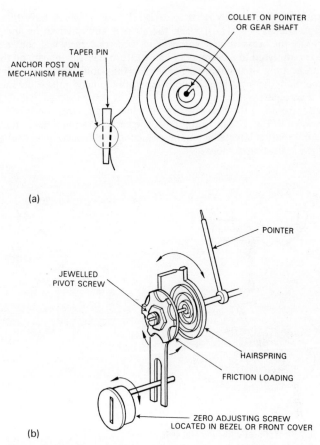

Fig 2.6 Hairsprings
(*a*) Method of attachment   (*b*) Method of zero adjustment

this and other changes in characteristics are not always desirable, and it therefore becomes necessary to take steps to neutralize those which, if unchecked, would introduce indication errors due solely to environmental temperature changes.

The methods adopted for *temperature compensation*, as it is called, are varied depending on the type of instrument to which they are applied. The oldest method of compensation is the one utilizing the bimetal strip principle and is applied to such instruments as airspeed indicators, altimeters, vertical speed indicators, and exhaust-gas temperature indicators.

Bimetal-strip Method
A bimetal strip, as the name implies, consists of two metals joined together at their interface to form a single strip. One of the metals is invar, a form of steel with a 36% nickel content and, a negligible coefficient of linear expansion,

while the other metal may be brass or steel both of which have high linear expansion coefficients. Thus, when the strip is subjected to an increase of temperature the brass or steel will expand, and conversely will contract when the strip is subjected to a decrease of temperature. The invar strip, on the other hand, on account of it having a negligible expansion coefficient, will always try to maintain the same length and being firmly joined to the other metal will cause the whole strip to bend.

An application of the bimetal-strip principle to a typical rod-type mechanism is shown in Fig 2.7 (*a*). In this case, the vertical ranging bar connected to the rocking shaft is bimetallic and bears against the arm coupled to the sector gear of the indicating element.

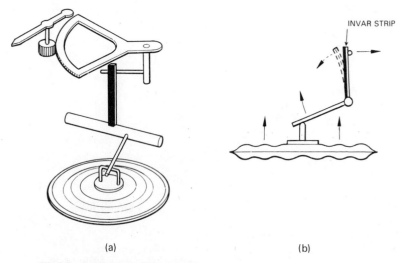

(a)   (b)

Fig 2.7 Application of bimetal strip

The principal effect which temperature changes have on this mechanism is expansion and contraction of the capsule, thus tending to make the indicating element overread or underread. For example, let us assume that the positions taken up by the mechanism elements are those obtaining when measuring a known quantity at the normal calibration temperature of 15°C, and that the temperature is gradually increased. The effect of the increase in temperature on the capsule material is to make it more flexible so that it will expand further to carry the ranging bar in the direction indicated by the solid arrows (Fig 2.7 (*b*)). As the ranging bar is in contact with the sector gear arm, the indicating element has the tendency to overread. However, the increase of temperature has a simultaneous effect on the ranging bar which, being a bimetal and on account of the position of the invar portion, will sag, or deflect in the direction indicated by the dotted arrow thus counteracting the capsule expansion and keeping the indicating element at a constant reading. When the temperature is decreased the capsule material "stiffens up" and contracts so that the indicating element tends to underread; as will be apparent from the diagram, a constant reading would be maintained by the bimetal ranging bar sagging or deflecting in the opposite direction.

In some instruments, for example exhaust-gas temperature indicators, indication errors can be introduced due to the effects of environmental temperature on the values of the electromotive force produced by a thermocouple system.

Although such errors ultimately result from changes in an electrical quantity, compensation can also be effected mechanically and by the application of the bimetal strip principle. As, however, the operation of the method is closely connected with the operating principles of thermo-electric instruments, we shall study it in detail at the appropriate stage.

Thermo-resistance Method

For temperature measurements in aircraft, the majority of instruments employed are of the electrical moving-coil type, and as the coil material is usually either copper or aluminium, changes of indicator temperature can cause changes in electrical resistance of the material. We shall be studying the fundamental principles of moving-coil instruments in a later chapter, but at this point we may note that, as they depend for their operation on electric current, which is governed by resistance, the effects of temperature can result in indication errors which necessitate compensation.

One of the compensation methods adopted utilizes a thermo-resistor or *thermistor* connected in the indicator circuit. A thermistor, which is composed of a mixture of metallic oxides, has a very large temperature coefficient of resistance which is usually negative; i.e. its resistance decreases with increases in temperature. Assuming that the temperature of the indicator increases, the current flowing through the indicator will be reduced because copper or aluminium will characteristically increase in resistance; the indicator would therefore tend to underread. The thermistor resistance will, on the other hand, decrease, so that for the same temperature change the resistance changes will balance out to maintain a constant current and therefore a constant indication of the quantity being measured.

Thermo-magnetic Shunt Method

As an alternative to the thermistor method of compensating for moving-coil resistance changes, some temperature measuring instruments utilize a device known as a *thermo-magnetic shunt*. This is a strip of nickel-iron alloy sensitive to temperature changes, which is clamped across the poles of the permanent magnet so that it diverts some of the airgap magnetic flux through itself.

As before, let us assume that the indicator temperature increases. The moving-coil resistance will increase thus opposing the current flowing through the coil, but, at the same time, the reluctance ("magnetic resistance") of the alloy strip will also increase so that less flux is diverted from the airgap. Since the deflecting torque exerted on a moving coil is proportional to the product of current and flux, the increased airgap flux counterbalances the reduction in current to maintain a constant torque and indicated reading. Depending on the size of the permanent magnet, a number of thermo-magnetic strips may be fitted to effect the required compensation.

SEALING OF INSTRUMENTS AGAINST ATMOSPHERIC EFFECTS

In pressurized aircraft, the internal atmospheric pressure conditions are increased to a value greater than that prevailing at the altitude at which the aircraft is flying. Consequently, instruments using external atmospheric pressure as a datum, for example altimeters, vertical speed indicators and airspeed indicators, are liable to inaccuracies in their readings should air at cabin pressure enter their cases. The cases are therefore sealed to withstand external pressures higher than those normally encountered under pressurized conditions. The external pressure against which sealing is effective is normally 15 lb/in^2.

Direct-reading pressure measuring instruments of the Bourdon tube, or capsule type, connected to a pressure source outside the pressure cabin, are also liable to errors. Such errors are corrected by using sealed cases and venting them to outside atmospheric pressure.

Many of the instruments in current use depend for their operation on sensitive electrical circuits and mechanisms which must be protected against the adverse effects of atmospheric temperature, pressure and humidity. This protection is afforded by filling the cases with an inert gas such as nitrogen or helium, and then hermetically sealing the cases.

QUESTIONS

2.1 What are the four principal elements which make up an instrument?

2.2 Define lever length and lever angle and state what effects they have on an instrument utilizing a lever mechanism.

2.3 (*a*) What are the essential differences between a lever mechanism and a rod mechanism? (*b*) State some typical applications to aircraft instruments.

2.4 What do you understand by the term "lost motion"? Describe the methods adopted for minimizing its effects.

2.5 Describe the methods by which instrument indications are automatically corrected for temperature variations. (SLAET)

3 Instrument Displays, Panels and Layouts

In flight, an aeroplane and its operating crew form a "man–machine" system loop, which, depending on the size and type of aircraft, may be fairly simple or very complex. The function of the crew within the loop is that of controller, and the extent of the control function is governed by the simplicity or otherwise of the machine as an integrated whole. For example, in manually flying an aeroplane, and manually initiating adjustments to essential systems, the controller's function is said to be a fully active one. If, on the other hand, the flight of an aeroplane and adjustments to essential systems are automatic in operation, then the controller's function becomes one of monitoring, with the possibility of reverting to the active function in the event of failure of systems.

Instruments, of course, play an extremely vital role in the control loop as they are the means of communicating data between systems and controller. Therefore, in order that a controller may obtain a maximum of control quality, and also to minimize the mental effort in interpreting data, it is necessary to pay the utmost regard to the content and form of the data display.

The most common forms of data display applied to aircraft instruments are (a) *quantitative*, in which the variable quantity being measured s presented in terms of a numerical value and by the relative position of a pointer or index, and (b) *qualitative*, in which the information is presented in symbolic or pictorial form.

QUANTITATIVE DISPLAYS

There are three principal methods by which information may be displayed: (i) the *circular scale*, or more familiarly, the "clock" type of scale, (ii) *straight scale*, and (iii) *digital*, or counter. Let us now consider these three methods in detail.

Circular Scale

This may be considered as the classical method of displaying information in quantitative form and is illustrated in Fig 3.1.

The *scale base*, or graduation circle, refers to the line, which may be actual or implied, running from end to end of the scale and from which the scale marks and line of travel of the pointer are defined.

Scale marks, or graduation marks, are the marks which constitute the scale of the instrument. For quantitative displays it is of extreme importance that the number of marks be chosen carefully in order to obtain quick and accurate interpretations of readings. If there are too few marks dividing the scale, vital information may be lost and reading errors may occur. If, on the other hand, there are too many marks, time will be wasted since speed of reading decreases as the number of markings increases. Moreover, an observer may get a spurious sense of accuracy if the number of scale marks makes it possible to read the scale accurately to, say, one unit (the smallest unit marked) when in actual fact the instrument has an inherent error causing it to be accurate to, say, two units. As far as quantitative-display aircraft instruments are concerned, a simple rule followed by manufacturers is to divide scales so that the marks represent units of 1, 2, or 5 or decimal multiples thereof. The sizes of the marks are also important and the general principle adopted is that the marks which are to be numbered are the largest while those in between are shorter and usually all of the same length.

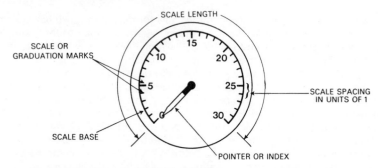

Fig 3.1 Circular scale quantitative display

Spacing of the marks is also of great importance, but since it is governed by physical laws related to the quantity to be measured, there cannot be complete uniformity between all quantitative displays. In general, however, we do find that they fall into two distinct groups, *linear* and *non-linear*; in other words, scales with marks evenly and non-evenly spaced. Typical examples are illustrated in Fig 3.2, from which it will also be noted that non-linear displays

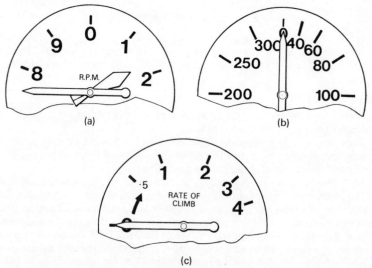

Fig 3.2 Linear and non-linear scales
(*a*) Linear   (*b*) Square-law   (*c*) Logarithmic

may be of the square-law or logarithmic-law type, the physical laws in this instance being related to airspeed and rate of altitude change respectively.

The sequence of numbering always increases in a clockwise direction thus conforming to what is termed the "visual expectation" of the observer. In an instrument having a centre zero, this rule would, of course, only apply to the positive scale. As in the case of marks, numbering is always in steps of 1, 2, or 5 or decimal multiples thereof. The numbers may be marked on the dial either inside or outside the scale base; the latter method is preferable since the numbers are not covered by the pointer during its travel over the scale.

The distance between the centres of the marks indicating the minimum and maximum values of the chosen range of measurement, and measured along the scale base, is called the *scale length*. Governing factors in the choice of scale length for a particular range are the size of the instrument, the accuracy with which it needs to be read, and the conditions under which it is to be observed. Under ideal conditions and purely from theoretical considerations, it has been calculated that the length of a scale designed for observing at a distance of 30 in and capable of being read to 1% of the total indicated quantity, should be about 2 in (regardless of its shape). This means that for a circular-scale instrument a 1 in diameter case would be sufficient. However, aircraft instruments must retain their legibility in conditions which at times may be far from ideal—conditions of changing light, vibrations imparted to the instrument panel, etc. In consequence, some degree of standardization of instrument case sizes was evolved, the utilization of such cases being dictated by the reading accuracy and the frequency at which observations are required. Instruments displaying information which is to be read accurately and at frequent intervals have scales about 7 in in length fitting into standard $3\frac{1}{4}$ in cases, while those requiring only occasional observation, or from which only approximate readings are required, have shorter scales and fit into smaller cases.

High-range Long-scale Displays

For the measurement of some quantities, for example, turbine-engine r.p.m., airspeed, and altitude, high measuring ranges are involved with the result that very long scales are required. This makes it difficult to display such quantities on single circular scales in standard-size cases, particularly in connection with the number and spacing of the marks. If a large number of marks are required their spacing might be too close to permit rapid reading, while, on the other hand, a reduction in the number of marks in order to open up the spacing, will also give rise to errors when interpreting values at points between scale marks.

Some of the displays developed as practical solutions to the difficulties encountered are illustrated in Fig 3.3. The display shown at (*a*) is perhaps the simplest way of accommodating a lengthy scale; by splitting it into two concentric scales the inner one is made a continuation of the outer. A single pointer driven through two revolutions can be used to register against both scales, but as it can also lead to too frequent mis-reading, a presentation by two interconnected pointers of different sizes is much better. A practical example of this presentation is to be found in some current designs of turbine-engine r.p.m. indicator. In this instance a large pointer rotates against the outer scale to indicate hundreds of r.p.m. and at the same time it rotates a smaller pointer against the inner scale indicating thousands of r.p.m.

In Fig 3.3 (*b*), we find a method which is employed in a certain type of airspeed indicator; in its basic concept it is similar to the one just described. In this design, however, a single pointer rotates against a circular scale and drives a second scale instead of a pointer. This rotating scale, which records hundreds of miles per hour as the pointer rotates through complete revolutions, is visible through an aperture in the main dial.

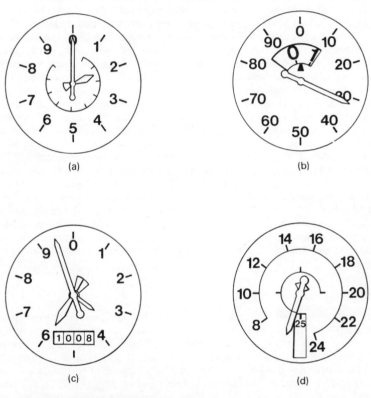

Fig 3.3 High-range long-scale displays

(*a*) Concentric scales (*c*) Common scale, triple pointers
(*b*) Fixed and rotating scales (*d*) Split pointer

A third method of presentation, shown at (*c*), is one in which three concentric pointers of different sizes register against a common scale. The application of this presentation has been confined mainly to altimeters, the large pointer indicating hundreds, the intermediate pointer thousands and the small pointer tens of thousands of feet. This method of presentation suffers several disadvantages the principal of which are that it takes too long to interpret a reading and gives rise to too frequent and too serious mis-reading.

Figure 3.3 (*d*), illustrates a comparatively recent presentation method applied to airspeed measurement. It will be noted that an outer and an inner scale are adopted and also what appears to be a single pointer. There are, however, two pointers which move together and register against the outer scale during their first revolution. When this has been completed, the tip of the longer pointer of the two is covered by a small plate and its movement beyond this point of the scale is arrested. The shorter pointer continues its movement to register against the inner scale.

Angle of Observation

Another factor which has an important bearing on the choice of the correct scale length and case size is the angle at which an instrument is to be observed. It is important because, even though it would be possible to utilize longer scales in the same relevant case sizes, the scale would be positioned so close to the outer edge of the dial plate that it would be obscured when observed at an angle. For this reason, a standard is also laid down that no part of an instrument should be obscured by the instrument case when observed at angles up to 30° from the normal. A method adopted by some manufacturers, which conforms to this standard, is the fitting of instrument mechanisms inside square cases.

When observing an instrument at an angle errors due to parallax are, of course, possible, the magnitude of such errors being governed principally by the angle at which the relevant part of its scale is observed, and also by the clearance distance between the pointer and dial plate. This problem like so many others in the instrument field has not gone unchallenged and the result is the "platform" scale designed for certan types of circular display instruments. As may be seen from Fig 3.4, the scale marks are set out on a circular platform which is secured to the main dial plate so that it is raised to the same level as the tip of the pointer.

Scale Range and Operating Range

A point quite often raised in connection with instrument scale lengths and ranges is they usually exceed that actually required for the operating range of the system with which the instrument is associated thus leaving part of the scale unused. At first sight this does appear to be somewhat wasteful, but an example will show that it helps in improving the accuracy with which readings may be observed.

Let us consider a fluid system in which the operating pressure range is say $0-36 \, lb/in^2$. It would be no problem to design a scale for the required pressure indicator which would be of a length equivalent to the system's operating range and also divided into a convenient number of parts as shown in Fig 3.5 (*a*). However, under certain operating conditions of the system concerned, it may be essential to monitor pressures having such values as 17 or $29 \, lb/in^2$ and to do this accurately in the shortest possible time is not very easy, as a second glance at the diagram will show. Let us now redesign the scale so that its length and range exceed the system's operating range and set out the scale marks according to the rule given on page 23. The result shown at (*b*) clearly indicates how easier it is to interpret the values we have considered it essential to monitor.

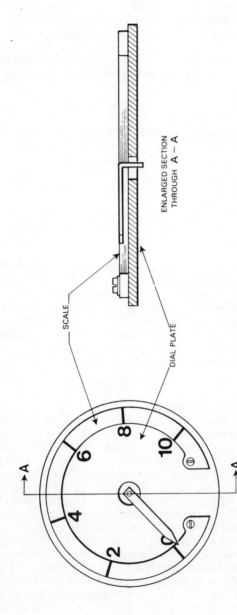

Fig 3.4 Platform scale

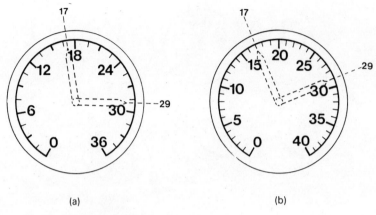

Fig 3.5 Reading accuracy
(a) Equal scale length and operating range
(b) Scale range exceeding operating range

Straight Scale

In addition to the circular scale presentation, a quantitative display may also be of the *straight scale* (vertical or horizontal) type. For the same reason that the sequence of numbering is given in a clockwise direction on a circular scale, so on a straight scale the sequence is from bottom to top or from left to right.

In the field of aircraft instruments there are very few applications of the straight scale and pointer displays, as they are not suitable for the monitoring of the majority of quantities to be measured. However, they do possess characteristics which can contribute to the saving of panel space and improved observational accuracy, particularly where the problems of grouping and monitoring a large number of engine instruments is concerned.

The development of these characteristics, and investigations into grouping and monitoring problems, have resulted in the practical application of another variation of the straight scale display. This is known as the *moving-tape* or "thermometer" display and is illustrated in Fig 3.6 as it would be applied to the measurement of two parameters vital to the operation of an aircraft powered by four turbojet engines.

Each display unit contains a servo-driven white tape in place of a pointer, which moves in a vertical plane and registers against a scale in a similar manner to the mercury column of a thermometer. As will be noted there is one display unit for each parameter, the scales being common to all four engines. By scanning across the ends of the tapes, or columns, a much quicker and more accurate evaluation of changes in engine performance can be obtained than from the classical circular scale and pointer display. This fact, and the fact that panel space can be considerably reduced, are also clearly evident from Fig. 3.6

Digital Display

A *digital* or *veeder-counter* type of display is one in which data are presented in the form of letters or numbers—*alpha-numeric display*, as it is technically termed. In aircraft instrument practice, the latter presentation is the most

30 Aircraft Instruments

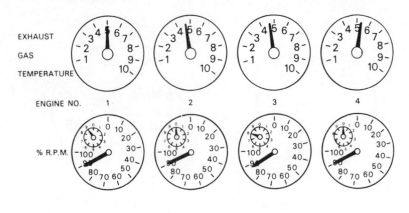

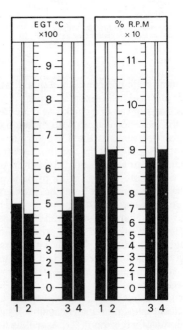

Engine No.	E.G.T. °C	% R.P.M.
1	500	89
2	470	90
3	480	88
4	520	90

Fig 3.6 Comparison between moving-tape and circular scale displays

common and a counter is generally to be found operating in combination with the circular type of display. A typical example of this is given in Fig 3.7 as applied to a current design of altimeter. In this particular application there are two counters: one presents a fixed pressure value which can be set mechanically by the pilot as and when required, and is known as a *static counter* display; the other is geared to the altimeter mechanism and automatically presents changes in altitude, and is therefore known as a *dynamic counter* display. It is of interest to note that the presentation of altitude data by means of a scale and counter is yet another method of solving the long-scale problem already discussed on page 25.

Fig 3.7   Application of digital display

Dual-indicator Displays

Dual-indicator displays are designed principally as a means of conserving panel space, particularly where the measurement of the various quantities related to engines is concerned. They are normally of two basic forms: one in which two separate indicators and scales are embodied in one case; and the other, also having two indicators in one case, but with the pointers registering against a common scale. Typical examples of display combinations are illustrated in Fig 3.8. Dual displays of the miniature type are usually confined to quantities which require to be observed only occasionally.

Coloured Displays

The use of colour in displays can add much to their value; not, of course from the artistic standpoint, but as a means of indicating specific operational ranges of the systems with which they are associated and to assist in making more rapid assessment of conditions prevailing when scanning the instruments.

Colour may be applied to scales in the form of sectors and arcs which embrace the number of scale marks appropriate to the required part of the range, and in the form of radial lines coinciding with appropriate individual

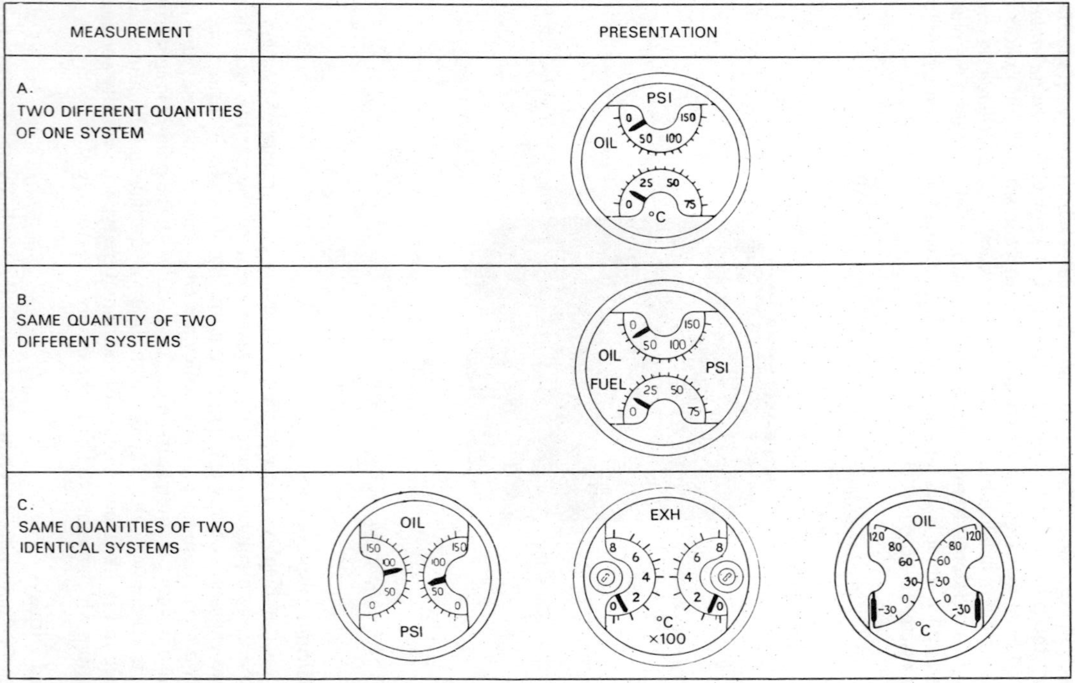
MEASUREMENT
PRESENTATION
A.
TWO DIFFERENT QUANTITIES OF ONE SYSTEM
PSI
OIL
0 50 100 150
0 25 50 75
°C
B.
SAME QUANTITY OF TWO DIFFERENT SYSTEMS
OIL
FUEL
PSI
0 50 100 150
0 25 50 75
C.
SAME QUANTITIES OF TWO IDENTICAL SYSTEMS
OIL
150 100 50 0
PSI
EXH
8 6 4 2 0
°C
×100
OIL
120 80 60 30 0 -30
°C

Fig 3.8 Examples of dual-indicator displays

scale marks. A typical example is illustrated in Fig 3.9. It is usual to find that coloured sectors are applied to those parts of a range in which it is sufficient to know that a certain condition has been reached rather than knowing actual quantitative values. The colours chosen may be red, yellow or green depending on the condition to be monitored. For example, in an aircraft oxygen system it may be necessary for the cylinders to be charged when the pressure has dropped to below, say, 500lb/in^2. The system pressure gauge would therefore have a red sector on its dial embracing the marks from 0 to 500; thus, if the pointer should register within this sector, this alone is sufficient indication that recharging is necessary and that it is only of secondary importance to know what the actual pressure is.

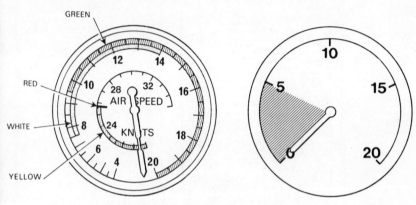

Fig 3.9 Use of colour in instrument displays
White arc 75–140   Yellow arc 225–255
Green arc 95–225   Red radial line 255

Arcs and radial lines are usually called *range markings*, their purpose being to define values at various points in the range of a scale which are related to specific operational ranges of an aircraft, its power plants and systems. The definitions of these marks are as follows:

RED *radial line*	Maximum and minimum limits
YELLOW *arc*	Take-off and precautionary ranges
GREEN *arc*	Normal operating range
RED *arc*	Range in which operation is prohibited

When applied to fuel quantity indicators, a RED *arc* indicates fuel which cannot be used safely in flight.

Airspeed indicator dials may also have an additional WHITE *arc*. This serves to indicate the airspeed range over which the aircraft landing flaps may be extended in the take-off, approach and landing configurations of the aircraft.

Range markings may vary for different aircraft types and are therefore added by the aircraft manufacturer prior to installation in their production aircraft.

It may often be found that markings are painted directly on the cover glasses of instruments—a method which is simpler since it does not require

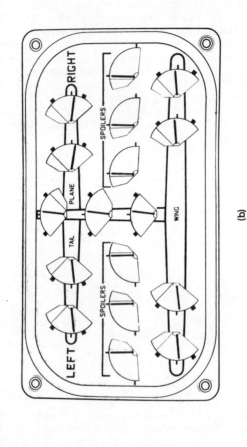

Fig 3.10 Qualitative displays
(a) Engine synchronizing (b) Position of flight control surfaces

removal of an instrument mechanism from its case. However, the precaution is always taken of painting a white index or register line half on the cover glass and half on the bezel to ensure correct alignment of the glass and the markings over the scale marks.

In addition to the foregoing applications, colour may also be used to facilitate the identification of instruments with the systems in which they are connected. For example, in one type of aircraft currently in service, triple hydraulic systems are employed, designated yellow system, green system and blue system, and in order to identify the pressure indicators of each system the scales are set out on dials painted in the relevant colours.

QUALITATIVE DISPLAYS

These are of a special type in which the information is presented in a symbolic or pictorial form to show the condition of a system, whether the value of an output is increasing or decreasing, the movment of a component and so on. Two typical examples are shown in Fig 3.10.

The synchroscope at (*a*) is used in conjunction with an r.p.m. indicating system of an aircraft having a multiple arrangement of propeller-type engines, and its pointers, which symbolize the propellers, only rotate to show the differences of speed between engines.

The display, shown at (*b*), is a good example of one indicating the movement of components; in this case, flight control surfaces, landing flaps, and air spoilers. The instrument contains seventeen separate electrical mechanisms, which on being actuated by transmitters, position symbolic indicating elements so as to appear at various angles behind apertures in the main dial.

DIRECTOR DISPLAYS

Director displays are those which are associated principally with flight attitude and navigational data (see Chapter 16), and presenting it in a manner which indicates to a pilot what control movements he must make either to correct any departure from a desired flight path, or to cause the aircraft to perform a specific manoeuvre. It is thus apparent that in the development of this type of display there must be a close relationship between the direction of control movements and the instrument pointer or symbolic-type indicating element; in other words, movements should be in the "natural" sense in order that the pilot may obey the "directives" or "demands" of the display.

Although flight director displays are of comparatively recent origin as specialized integrated instrument systems of present-day aircraft, in concept they are not new. The gyro horizon (see page 119) which has been in use for many years utilizes in basic form a director display of an aircraft's pitch and bank attitude. In this instrument there are three elements making up the display: a pointer registering against a bank-angle scale, an element symbolizing the aircraft, and an element symbolizing the natural horizon. Both the bank pointer and natural horizon symbol are stabilized by a gyroscope. As the instrument is designed for the display of attitude angles, and as also one of the symbolic elements can move with respect to the other, then it has two reference axes, that of the case which is fixed with respect to the aircraft, and that of the moving element. Assuming that the aircraft's pitch attitude changes to bring

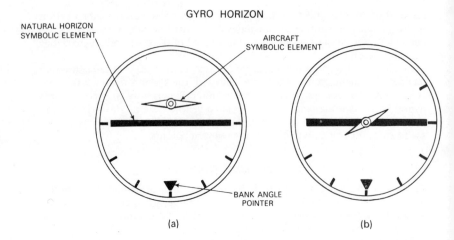

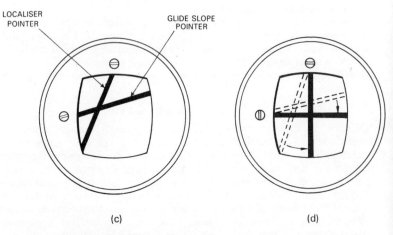

Fig 3.11 Basic examples of director display
(a) "Fly down" directive
(b) "Bank right" directive
(c) "Fly left" and "Fly up" directive
(d) Response matches directive

the nose up, then the horizon display will be as shown in Fig 3.11 (a), thus directing or demanding the pilot to "get the nose down." Similarly, if the bank attitude should change whereby the left wing goes down, then the horizon display would be as shown at (b), directing or demanding the pilot to "bank the aircraft to the right." In both cases, the demands would be satisfied by the pilot moving his controls in the natural sense.

Another example of a basic director display is that utilized in the indicator forming part of the *Instrument Landing System* (ILS); this is a system which aids a pilot in maintaining the correct position of his aircraft during the approach to land on an airport runway. Two radio signal beams are transmitted from the ground; one beam is in the vertical plane and at an angle to the runway to establish the correct approach or glide slope angle; while the other, known as the localizer, is in the horizontal plane; both are lined up with the runway centre-line.

A receiver on board the aircraft receives the signals and transmits them to a cross-pointer type of instrument on the main instrument panel. Meters within the instrument are monitored by the glide slope signals and localizer signals, and respectively control horizontal and vertical pointers.

When the aircraft is on the approach to land and is, say, below the glide slope beam, the horizontal pointer of the instrument will be deflected upwards as shown in Fig 3.11 (*c*). Thus, the pilot is directed to "fly the aircraft up" in order to intercept the beam. Similarly, if the aircraft is to the right of the localizer beam the vertical pointer will be deflected to the left thus directing the pilot to "fly the aircraft left." As the pilot responds to the instrument's directives the pointers move back until they are crossed at the centre of the instrument dial, indicating that the aircraft is in the correct approach position for landing.

It will be apparent from the diagram that as the aircraft is manoeuvred in response to demands, the pointer movements are contrary to the "natural" sense requirements; for example, in responding to the demand "fly left" the vertical pointer will move to the right. However, in turning to the left the bank attitude of the aircraft will change into the direction of the turn, that is, the left wing will go down, and as the gyro horizon will indicate this directly, then by monitoring this instrument the pilot can cross-check that he is putting the aircraft into the correct attitude in responding to the demands of the ILS indicator.

Although the foregoing methods of displaying "demand and response" data are satisfactory, they fall very short in meeting the requirements for precision attitude flying particularly during the approach and landing phases of a flight. As we have just noted, a pilot must interpret the demands of one instrument, make corrections, and note the responses by observing this instrument and cross-checking on a second. If demands and appropriate responses required to match them can be displayed on a single instrument, with all movements of pointers or symbolic indicating elements in the natural sense, then a pilot's task will be that much easier. As an illustration of this line of approach to an integrated director display, let us consider the one shown in Fig 3.12, which is, in fact, the display of a Smith's *Flight System Director Horizon*.

In appearance, the presentation resembles that of a normal gyro horizon; however, there are two additional indicating elements, one for showing pitch demands, and the other for showing bank or turn demands. If the aircraft is flying straight and level and a "fly left" demand signal, say, is received, the azimuth director pointer will move to the left and the display will then be as shown at (*b*). In responding to the demand the pilot will bank the aircraft to the left, thus turning it in the same direction, until the ring-sight type of bank pointer circumscribes the director pointer. The display will then be as shown at (*c*), clearly indicating the attitude of the aircraft after matching of response to demand. In the case of signals demanding a change in pitch, a "fly up" demand for example, the displays would be as shown at (*d*) and (*e*).

38 *Aircraft Instruments*

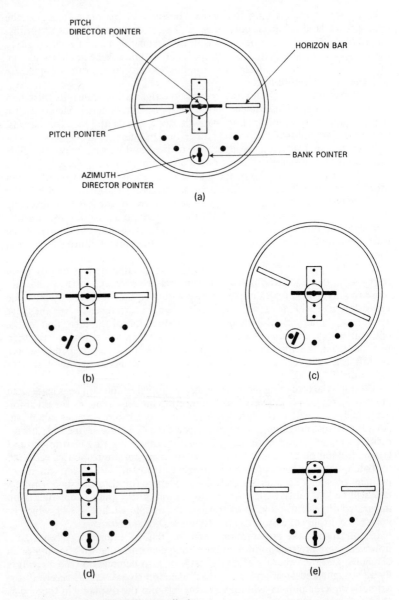

Fig 3.12 Integrated director display
(a) Indicating elements
(b) "Fly left" demand
(c) Response matches demand by banking aircraft to the left
(d) "Fly up" demand
(e) Response matches demand by climbing the aircraft

HEAD-UP DISPLAYS

From the descriptions thus far given of the various instrument displays, we have gained some idea of the development approach to the problem of presenting data which is to be quickly and accurately assimilated. The simplicity or otherwise of assimilation is dictated by the number of instruments involved, and by the amount of work and instrument monitoring sequences to be performed by a pilot during the various phases of flight. In the critical approach and landing phase, a pilot must transfer his attention more frequently from the instruments to references outside the aircraft, and back again; a transition process which is time-consuming and fatiguing as a result of constant re-focusing of the eyes.

A method of alleviating these problems has therefore been developed in which vital flight data is presented at the same level as the pilot's line-of-sight when viewing external references, i.e. when he is maintaining a "head-up" position. The principle of the method is to display the data on the face of a special cathode-ray tube and to project it optically as a composite symbolic image on to a transparent reflector plate, or directly on the windscreen. The components of a typical head-up display system are shown in Fig 3.13. The amount of data required is governed by the requirements of the various flight phases and operational role of an aircraft, i.e. civil or military, but the four parameters shown are basic. The data is transmitted from a data computer unit to the cathode-ray tube the presentation of which is projected by the optical system to infinity. It will be noted that the attitude presentation resembles that of a normal gyro horizon, and also that airspeed and altitude are presented by markers which register against linear horizontal and vertical scales. The length, or range, of the scales is determined by operational requirements, but normally they only cover narrow bands of airspeed and altitude information. This helps to reduce irrelevant markings and the time taken to read and interpret the information presented.

INSTRUMENT PANELS AND LAYOUTS

All instruments essential to the operation of an aircraft are accommodated on special panels the number and distribution of which vary in accordance with the number of instruments, the size of aircraft and cockpit layout. A main instrument panel positioned in front of pilots is a feature common to all types of aircraft, since it is mandatory for the primary flight instruments to be installed within the pilots' normal line of vision. Typical positions of other panels are: overhead, at the side, and on a control pedestal located centrally between the pilots (Fig 3.14).

Panels are invariably of light alloy of sufficient strength and rigidity to accommodate the required number of instruments, and are attached to the appropriate parts of the cockpit structure. The attachment methods adopted vary, but all should conform to the requirement that a panel or an individual instrument should be easily installed and removed.

Main instrument panels which may be of the single-unit type or made up of two or three sub-panel assemblies, are supported on shockproof mountings since they accommodate the flight instruments and their sensitive mechanisms. The number, size and disposition of shockproof mountings required are governed by the size of panel and distribution of the total weight.

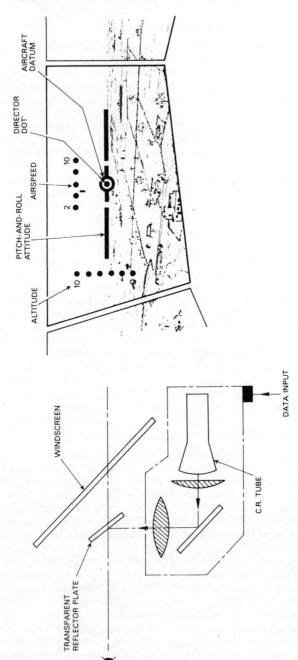

Fig 3.13 Head-up display system

Fig 3.14 Location of instrument panels in a turbojet airliner

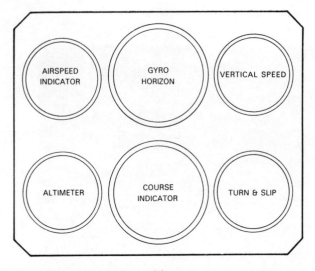

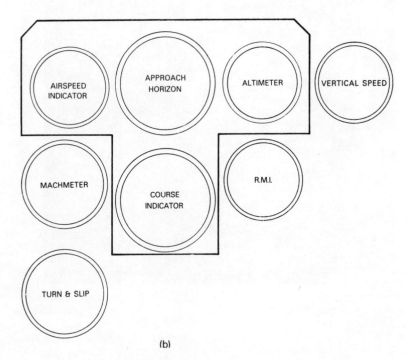

Fig 3.15 Flight instrument grouping
(a) "Basic six"   (b) "Basic T"

All panels are normally mounted in the vertical position, although in some current aircraft types the practice of sloping main instrument panels forward at about 15° from the vertical is adopted to minimize parallax errors.

Instrument and all other control panels which for many years were painted black, are now invariably finished in matt grey, a colour which apart from its "softer" effects provides a far better contrasting background for the instrument dials and thus contributes to easier identification.

INSTRUMENT GROUPING

Flight Instruments

Basically there are six flight instruments whose indications are so co-ordinated as to create a "picture" of an aircraft's flight condition and required control movements; they are, airspeed indicator, altimeter, gyro horizon, direction indicator, vertical speed indicator and turn-and-bank indicator. It is therefore most important for these instruments to be properly grouped to maintain co-ordination and to assist a pilot to observe them with the minimum of effort.

The first real attempt at establishing a standard method of grouping was the "blind flying panel" or "basic six" layout shown in Fig 3.15 (a). The gyro horizon occupies the top centre position, and since it provides positive and direct indications of attitude, and attitude changes in the pitching and rolling planes, it is utilized as the master instrument. As control of airspeed and altitude are directly related to attitude, the airspeed indicator, altimeter and vertical speed indicator flank the gyro horizon and support the interpretation of pitch attitude. Changes in direction are initiated by banking an aircraft, and the degree of heading change is obtained from the direction indicator; this instrument therefore supports the interpretation of roll attitude and is positioned directly below the gyro horizon. The turn-and-bank indicator serves as a secondary reference instrument for heading changes, so it too supports the interpretation of roll attitude.

With the development and introduction of new types of aircraft, flight instruments and integrated instrument systems, it became necessary to review the functions of certain instruments and their relative positions within the group. As a result a more effective and standardized grouping has now been adopted; this is known as the "basic T" and is shown in Fig 3.15 (b). The theory behind this method is that it constitutes a system by which various items of related flight information can be placed in certain standard locations in all instrument panels regardless of type or make of instrument used. In this manner, advantage can be taken of integrated instruments which display more than one item of flight information.

It will be noted that there are now four "key" instruments, airspeed indicator, pitch and roll attitude indicator, an altimeter forming the horizontal bar of the "T," and the direction indicator forming the vertical bar. As far as the positions flanking the direction indicator are concerned, they are taken by other but less specifically essential flight instruments, and there is a certain degree of freedom in the choice of function. From Fig 3.15 it can be seen, for example, that a Machmeter and a radio magnetic indicator can take precedence over a turn-and-bank indicator and a vertical-speed indicator.

Border lines are usually painted on the panel around the flight instrument groups. These are referred to as "mental focus lines," their purpose being to

Fig 3.16 Instrument grouping at a flight engineer's station

assist pilots in focusing their attention on and mentally recording the position of instruments within the groups.

Power-plant Instruments

The specific grouping of instruments required for the operation of power plants is governed primarily by the type of power plant, the size of the aircraft and therefore the space available. In a single-engined aircraft, this does not present much of a problem since the small number of instruments may flank the pilot's flight instruments thus keeping them within a small "scanning range."

The problem is more acute in multi-engined aircraft; duplication of power plants means duplication of their essential instruments. For twin-engined aircraft, and for certain medium-size four-engined aircraft, the practice is to group the instruments at the centre of the main instrument panel and between the two groups of flight instruments.

In the larger and more complex of the four-engined aircraft, a flight engineer's station is provided in the crew compartment and all the power plant instruments are grouped on the control panels at this station (Fig 3.16). Those instruments measuring parameters required to be known by a pilot during take-off, cruising and landing, e.g. r.p.m. and turbine temperature, are duplicated on the main instrument panel.

The positions of the instruments in the power plant group are arranged so that those relating to each power plant correspond to the power plant positions as seen in plan view. It will be apparent from the layout of Fig 3.17 that by scanning a row of instruments a pilot or engineer can easily compare the readings of a given parameter, and by scanning a column of instruments can assess the overall performance pattern of a particular power plant. Another advantage of this grouping method is that all the instruments for one power plant are more easily associated with the controls for that power plant.

METHODS OF MOUNTING INSTRUMENTS

The two methods most commonly used for the panel mounting of instruments are the flanged case method, and the clamp method. The flanged case method requires the use of screws inserted into locking nuts which, in some instruments, are fitted integral with the flange.

Since flanged-type indicators are normally mounted from the rear of the panel, it is clear that the integrally fixed locking nuts provide for much quicker mounting of an instrument and overcome the frustration of trying to locate a screw in the ever-elusive nut!

As a result of the development of the hermetic sealing technique for instruments, the cases of certain types are flangeless, permitting them to be mounted from the front of the instrument panel. In order to secure the instruments special clamps are provided at each cut-out location. The clamps are shaped to suit the type of case, i.e. circular or square, and they are fixed on the rear face of the panel so that when an instrument is in position it is located inside the clamp. Clamping of the instrument is effected by rotating adjusting screws which draw the clamp bands tightly around the case.

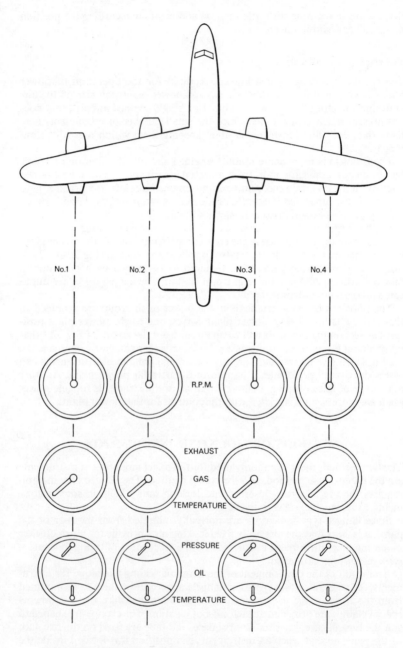

Fig 3.17 Power plant instrument grouping

MAGNETIC INDICATORS AND "FLOW LINES"

In the larger types of aircraft numerous valves, actuators and similar devices are used in many of their systems to obtain the desired control of system operation; for example, in a fuel system, actuators position valves which permit the supply of fuel from the main tanks to the engines and also crossfeed the fuel supply.

All such devices are, in the majority of cases, electrically operated and controlled by switches on the appropriate systems panel, and to confirm the completion of movement of the device an indicating system is necessary.

The indicating system could be in the form of a scale and pointer type of instrument or an indicator light. Both methods, however, have certain disadvantages. The use of an instrument is rather space-consuming, particularly where a number of actuating devices are involved, and unless it is essential for a pilot or systems engineer to know exactly the position of a device at any one time, instruments are uneconomical. Indicator lights are of course simpler, cheaper and consume less power, but the liability of their filaments to failure without warning constitutes a hazard, particularly in the case where "light out" is intended to indicate a safe condition of a system. Furthermore, in systems requiring a series of constant indications of prevailing conditions, constantly illuminated lamps can lead to confusion, misinterpretation (frustration also!) on the part of the pilot or engineer.

Therefore to enhance the reliability of indication, indicators containing small electromagnets operating a shutter or similar moving element are installed on the systems panels of many present-day aircraft.

In its simplest form (Fig 3.18 (a)) a *magnetic indicator* is of the two-position type comprising a ball pivoted on its axis and spring-returned to the "off" position. A ferrous armature, embedded in the ball, is attracted by the electromagnet when energized, and rotates the ball through 150° to present a different picture in the window. The picture can be either of the line-diagram type, representing the flow of fluid in a system (see Fig 3.19), or of the instructive type presenting such legends as OFF, ON, OPEN, CLOSE.

Figure 3.18 (b) shows a development of the basic indicator. It incorporates a second electromagnet which provides for three alternative indicating positions. The armature is pivoted centrally above the two magnets and can be attracted by either of them. Under the influence of magnetic attraction the armature tilts and its actuating arm slides the rack horizontally to rotate the pinions fixed to the ends of prisms. The prisms will then be rotated through 120° to present a new pattern in the window. When the rack moves from the central "rest" position, one arm of the hairpin type centring spring, located in a slot in the rack, will be loaded. Thus, if the electromagnet is de-energized, the spring will return to mid-position, rotating the pinions and prisms back to the "off" condition in the window.

The pictorial representation offered by these indicators is further improved by the painting of "flow lines" on the appropriate panels so that they interconnect the indicators with the system control switches and essential indicating instruments and warning lights.

A typical application of magnetic indicators and flow lines is shown in Fig 3.19.

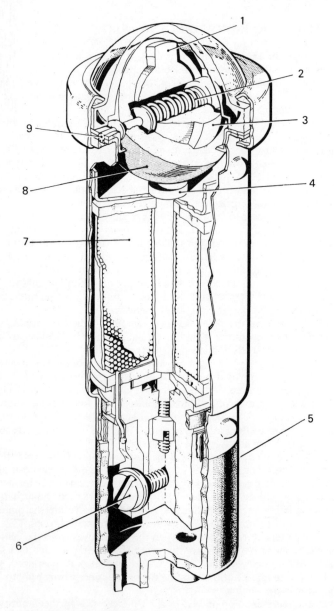

Fig 3.18 Typical magnetic indicators
(a) Section through two-position indicator
1 Armature
2 Return spring
3 Balance weight
4 Magnet assembly
5 Moisture-proof grommet
6 Terminal
7 Coil
8 Plastic ball
9 Spindle

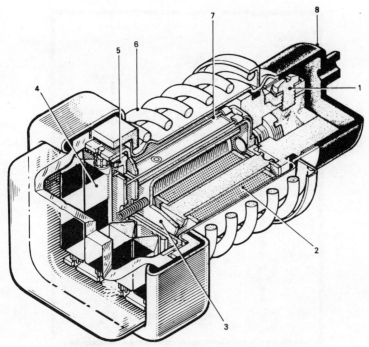

Fig 3.18 (contd.)

(b) Prism-type indicator
1 Terminal 5 Rack
2 Coils 6 Mounting spring
3 Armature 7 Control spring
4 Prisms 8 Cable grommet

ILLUMINATION OF INSTRUMENTS AND INSTRUMENT PANELS

When flying an aircraft at night, or under adverse conditions of visibility, a pilot is dependent on instruments to a much greater extent than he is when flying in daylight under good visibility conditions, and so the ability to observe their readings accurately assumes greater importance. For example, at night, the pilot's attention is more frequently divided between the observation of instruments and objects outside the aircraft, and this of course results in additional occular and general fatigue being imposed on him. Adequate illumination of instruments and the panels to which they are fitted is therefore an essential requirement.

The colour chosen for lighting systems has normally been red since this is considered to have the least effect on what is termed the "darkness adaptation characteristic" of the eyes. As a result of investigations and tests carried out in recent years, however, it would appear that white light has less effect, and this is now being used in some current types of aircraft.

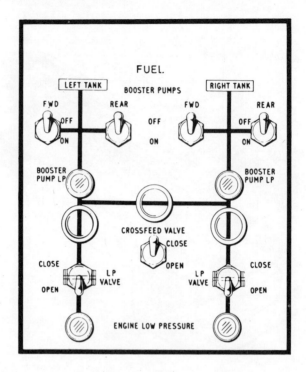

Fig 3.19 Application of magnetic indicators and flow lines

Fluorescent Dial Markings and Ultra-violet Flood Lighting

This method of instrument illumination is an early one and depends for its operation on the light emitted by certain substances when irradiated by ultra-violet rays.

For aircraft in which such a method is to be adopted, manufacturers produce instruments having scale marks and numerals treated with a special fluorescent compound. Only a selected few of the marks and numerals are so treated; the principal reasons for this are that it reduces the number of bright markings which could cause observation errors, and that the markings selected are considered as being only those an experienced pilot requires when flying at night or under adverse visibility conditions.

The lamps are fluorescent tubes constructed of special glass and coated on the inside with a fluorescent powder. Electrical power for energizing the lamps is derived from the aircraft's 115 V 400 Hz supply system and is controlled by switches and rheostats for varying the lighting intensity.

The number of lamps and their positions depend on the size and layout of the aircraft cockpit, but for the illumination of a main instrument panel three are generally sufficient: one at each end of the panel and one at the centre.

Pillar and Bridge Lighting

Pillar lighting, so called from the method of construction and attachment of the lamp, provides illumination for individual instruments and controls on the various cockpit panels. A typical assembly, shown in Fig 3.20(*a*), consists of a miniature centre-contact filament-lamp inside a housing, which is a push fit into the body of the assembly. The body is threaded externally for attachment to the panel and has a hole running through its length to accommodate a cable which connects the positive supply to the centre contact. The circuit through the lamp is completed by a ground tag to connect to the negative cable.

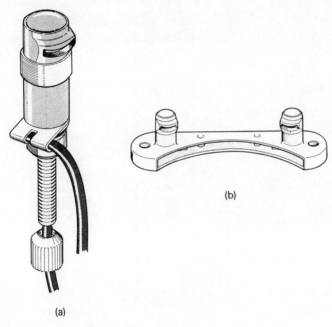

Fig 3.20   Pillar light assemblies

Light is distributed through a red filter and an aperture in the lamp housing. The shape of the aperture distributes a sector of light which extends downwards over an arc of approximately 90° to a depth slightly less than 2 in from the mounting point.

The bridge-type of lighting (Fig 3.20 (*b*)) is a multi-lamp development of the individual pillar lamp already described. Two or four lamps are fitted to a bridge structure designed to fit over a variety of the standardized instrument cases. The bridge fitting is composed of two light-alloy pressings secured together by rivets and spacers, and carrying the requisite number of centre contact assemblies above which the lamp housings are mounted. Wiring arrangements provide for two separate supplies to the lamps thus ensuring that loss of illumination cannot occur as a result of failure of one circuit.

Wedge-type Lighting

This method of instrument lighting derives its name from the shape of the two portions which together make up the instrument cover glass. It relies for its operation upon the physical law that the angle at which light leaves a reflecting surface equals the angle at which it strikes that surface.

The two wedges are mounted opposite to each other and with a narrow airspace separating them, as shown in Fig 3.21. Light is introduced into wedge A

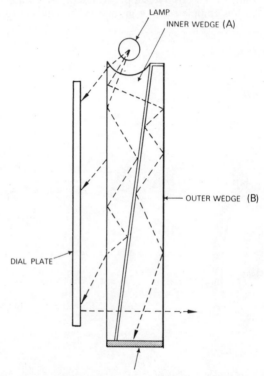

Fig 3.21 Wedge-type lighting

from two 6 V lamps set into recesses in its wide end. A certain amount of light passes directly through this wedge and onto the face of the dial while the remainder is reflected back into the wedge by its polished surfaces. The angle at which the light rays strike the wedge surfaces governs the amount of light reflected; the lower the angle, the more light reflected.

The double wedge mechanically changes the angle at which the light rays strike one of the reflecting surfaces of each wedge, thus distributing the light evenly across the dial and also limiting the amount of light given off by the instrument. Since the source of light is a radial one, the initial angle of some light rays with respect to the polished surfaces of wedge A is less than that of the others. The low-angle light rays progress further down the wedge before they leave and spread light across the entire dial.

Light escaping into wedge B is confronted with constantly decreasing angles and this has the effect of trapping the light within the wedge and directing it to its wide end. Absorption of light reflected into the wide end of wedge B is ensured by painting its outer part black.

Illuminated Instrument Panel

A unique method of illuminating instrument panels, the Thorn Plasteck system, is shown in Fig 3.22.

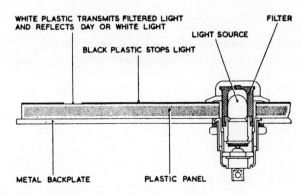

Fig 3.22   Illuminated instrument panel

In this system, a relatively thick sheet of acrylic plastic is faced on both its main surfaces by a thin sheet of translucent white plastic. Over this is laid a second thin sheet of black opaque plastic and the whole assembly is then bonded together to form a homogeneous panel. Cut-outs, corresponding to the locations of the instruments to be mounted on the panel, are made in the plastic panel which is fitted as an overlay.

Illumination is provided by small lampholders incorporating red filters and built into the panel at carefully selected positions. Thus, red light is transmitted through the acrylic sheet to the inside edge of the cut-outs and is spread evenly over the dials of the instruments.

A further advantage of this system is that numerals, operational data and other superscriptions, can be made directly on the panel by engraving through the black outer layer without piercing the white layer. In daylight, therefore, all superscriptions show clearly and vividly against the matt black surface of the panel, while at night they are back-illuminated in red.

QUESTIONS

3.1   Describe two of the methods adopted for the display of indications related to high-range measurements.

3.2   What is the purpose of a "platform scale"? Describe its arrangement.

3.3   Name some of the aircraft instruments to which a digital counter display is applied.

54 *Aircraft Instruments*

3.4 What is the significance of coloured markings applied to the dials of certain instruments?

3.5 If it is necessary to apply coloured markings to the cover glass of an instrument, what precautions must be taken?

3.6 What types of display would you associate with the following instruments: (*a*) a synchroscope, (*b*) an altimeter, (*c*) a gyro horizon?

3.7 What do you understand by the term head-up display? With the aid of diagrams describe how required basic flight data is displayed to a pilot.

3.8 Describe the "basic T" method of grouping flight instruments.

3.9 Describe the method by which a hermetically-sealed instrument is mounted on a panel.

3.10 Describe one of the methods of illuminating instrument dials.

3.11 What is the function of a magnetic indicator? Explain its operating principle.

4 Pitot-static Instruments and Systems

PITOT-STATIC SYSTEM

The *pitot-static* system of an aircraft is a system in which total pressure created by the forward motion of the aircraft and the static pressure of the atmosphere surrounding it are sensed and measured in terms of speed, altitude and rate of change of altitude (vertical speed).

In its basic form the system consists of a pitot-static tube, or pressure head, the three primary flight instruments—airspeed indicator, altimeter and vertical speed indicator—and pipelines and drains, interconnected as shown diagrammatically in Fig 4.1.

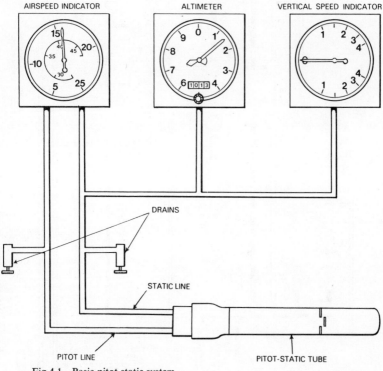

Fig 4.1 Basic pitot-static system

The complexity of a pitot-static system depends primarily upon the type and size of aircraft, the number of locations at which primary flight instrument data is required, and the types of instrument. The point about complexity is clearly borne out by comparing Figs 4.1 and 4.2.

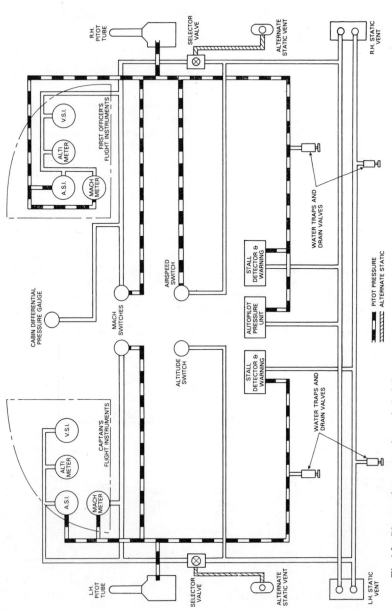

Fig 4.2 Pitot-static system of a typical airliner

Pitot-static Instruments and Systems

Sensing of the total, or pitot pressure, and of the static pressure is effected by the pressure head, which is suitably located in the airstream and transmits these pressures to the instruments. The pressure head (Fig 4.3) is shown in its simplest and original form to serve as the basis for understanding pitot and static pressure measurement.

It consists of two forward-facing tubes positioned parallel to each other and in a vertical plane. One of the tubes, the pitot tube, is open at its forward end to receive the total air pressure resulting from the aircraft's forward movement, while the other, the static tube, is closed at its forward end but has a series of small holes drilled circumferentially at a calculated distance from the forward end through which the undisturbed air at prevailing atmospheric pressure is admitted. Pressures are transmitted to the instruments through pipelines connected to each tube.

Pitot Pressure

This may be defined as the additional pressure produced on a surface when a flowing fluid is brought to rest, or stagnation, at the surface.

Let us consider a pitot tube placed in a fluid with its open end facing upstream as shown in Fig 4.4. When the fluid flows at a certain velocity V over the tube it will be brought to rest at the tube entry and this point is known as the *stagnation point*. If the fluid is an ideal one, i.e. is not viscous, then the total energy is equal to the sum of the potential energy, the kinetic energy and pressure energy, and remains constant. In connection with a pressure head, however, the potential energy is neglected thus leaving the sum of the remaining two terms as the constant. Now, in coming to rest at the stagnation point, kinetic energy of the fluid is converted into pressure energy. This means that work must be done by the mass of liquid and this raises an equal volume of the fluid above the level of the fluid stream.

If the mass of the fluid above this level is m pounds then the work done in raising it through a height h feet is given by

Work done $= mgh$ foot-pounds

where g is the acceleration due to gravity. In the British Isles at sea-level, $g = 32 \cdot 2 \text{ ft/s}^2 = 9 \cdot 81 \text{ m/s}^2$.

The work done is also equal to the product of the ratio of mass to density (ρ) and pressure (p):

Work done $= \dfrac{m}{\rho} p$

The kinetic energy of a mass m before being brought to rest is equal to $\tfrac{1}{2}mV^2$, where V is the speed, and since this is converted into pressure energy,

$$\frac{m}{\rho} p = \tfrac{1}{2}mV^2$$

Therefore

$$p = \tfrac{1}{2}\rho V^2$$

and is additional to the static pressure in the region of fluid flow.

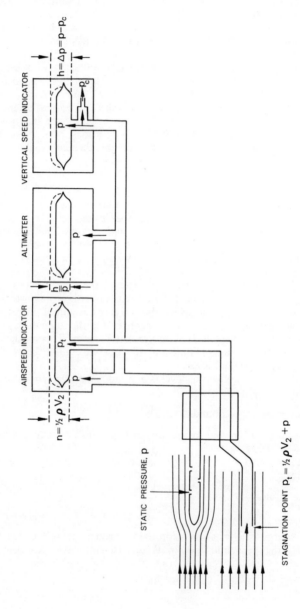

Fig 4.3 Sensing and transmission of pitot and static pressures

The factor $\frac{1}{2}$ assumes that the fluid is an ideal one and so does not take into account the fact that the shape of a body subjected to fluid flow may not bring the fluid to rest at the stagnation point. However, this coefficient is determined by experiment and for pressure heads it has been found that its value corresponds almost exactly to the theoretical one.

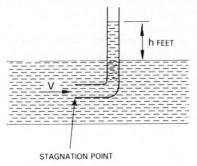

Fig 4.4 Pitot pressure

The $\frac{1}{2}\rho V^2$ law, as it is usually called in connection with airspeed measurement, does not allow for the effects of compressibility of air and so other factors must be introduced. These effects and the modified law will be covered in the section dealing with airspeed indicators.

Pressure Heads

As a result of aerodynamic "cleaning up" of aircraft it became necessary for changes to be made in the design of pressure heads in order to measure the pressures more accurately and without causing serious airflow disturbances around the tubes. A further requirement, and one which could not be met satisfactorily with pressure heads employing separate tubes, was the provision of a heating system to prevent the tubes from icing up when flying in icing conditions. The required design changes resulted in the arrangement shown in basic form in Fig 4.5.

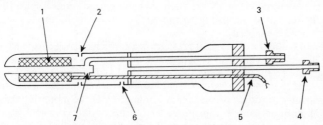

Fig 4.5 Basic form of pressure head

1 Heating element
2 Static slots
3 Pitot tube connection
4 Static tube connection
5 Heater element cable
6 External drain hole
7 Pitot tube drain hole

The tubes are mounted concentrically, the pitot tube being inside the static tube, which also forms the casing. Static pressure is admitted through either slots or small holes around the casing. The pressures are transmitted from their respective tubes by means of metal pipes which may extend to the rear of the pressure head, or at right angles, depending on whether it is to be mounted at the leading edge of a wing, under a wing, or at the side of a fuselage. Locations of pressure heads will be covered in more detail under the heading of "Pressure Error."

A chamber is normally formed between the static slots or holes and the pipe connection to smooth out any turbulent air flowing into the slots, which might occur when the complete tube is yawed, before transmitting it to the instruments.

The heating element is fitted around the pitot tube, or in some designs around the inner circumference of the outer casing, and in such a position within the casing that the maximum heating effect is obtained at the points

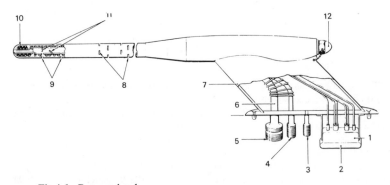

Fig 4.6   Pressure head

1	Terminal block	5	Rear-slot static connection	9	Drain holes
2	Cover	6	Rear heater	10	Nose heaters
3	Pitot pressure connection	7	Mast	11	Baffles
4	Front-slot static connection	8	Static slots	12	Mast drain-screw

where ice build-up is most likely to occur. The temperature/resistance characteristics of some elements are such that the current consumption is automatically regulated according to the temperature conditions to which the pressure head is exposed.

Figure 4.6 illustrates a type of pressure head which is fitted to an airliner currently in service. The head is supported on a mast which is secured to the fuselage skin by means of a suitably angled and profiled mounting flange. The tube assembly incorporates two sets of static slots connected separately and independently to static pressure pipes terminating at the mounting flange (see page 67). Pitot pressure is transmitted to the appropriate connecting union via a sealed chamber formed by the mast. To prevent the entry of water and other foreign matter, the pitot pressure tube is provided with baffles. Drain holes forward and to the rear of the baffles prevent the accumulation of moisture within the tube assembly. Water which might condense and accumulate in the mast can be drained by removing a drain screw located in the position shown. It should be noted that in addition to the pitot tube and casing, the mast is also protected against ice formation by its own heating element.

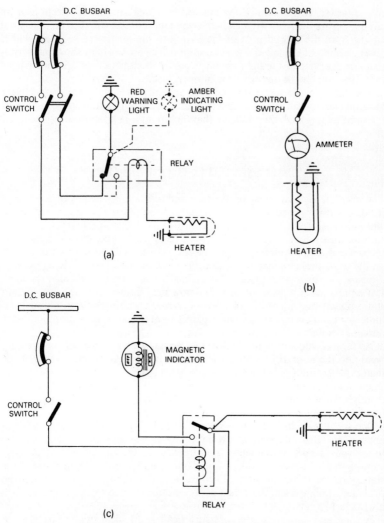

Fig 4.7 Typical heating circuit arrangements
(a) Light and relay   (b) Ammeter   (c) Magnetic indicator and relay

HEATING CIRCUIT ARRANGEMENTS

The direct current for heating is controlled by a switch located on a cockpit control panel, and it is usual to provide some form of indication of whether or not the circuit is functioning correctly. Three typical circuit arrangements are given in Fig 4.7.

In the arrangement shown at (a) the control switch, when in the "on" position, allows current to flow to the heater via the coil of a relay which will be energized when there is continuity between the switch and the grounded side of the heater. If a failure of the heater, or a break in another section of its circuit, occurs the relay will de-energize and its contacts will then complete the circuit from the second pole of the switch to illuminate the red light which gives warning of the failed circuit condition. The broken lines show an alternative arrangement of the light circuit whereby illumination of an amber light indicates that the heater circuit is in operation.

The indicating devices employed in arrangements (*b*) and (*c*) are respectively an ammeter and an on-off magnetic indicator. The function in both cases is the same as at (*a*), i.e. they indicate that the circuit is in operation.

PRESSURE (POSITION) ERROR

The accurate measurement of airspeed and altitude by means of a pressure head, has always presented two main difficulties: one, to design a head which will not cause any disturbance to the airflow over it; and the other, to find a suitable location on an aircraft where the head will not be affected by air disturbances due to the aircraft itself. The effects of such disturbances are greatest on the static-pressure measuring section of a pitot-static system giving rise to a *pressure*, or *position, error*, which is defined as the amount by which the local static pressure at a given point in the flow field differs from the free-stream static pressure. As a result of pressure error, an altimeter and an airspeed indicator can develop positive or negative errors. The vertical speed indicator remains unaffected.

As far as airflow over the head is concerned, we may consider the pressure head and the aircraft to which it is fitted as being alike because some of the factors determining air flow are: shape, size, speed and angle of attack. The shape and size of the head are dictated by the speed at which it is moved through the air; a large-diameter casing, for example, can present too great a frontal area which at very high speeds can initiate the build-up of a shock wave which will break down the flow over the head. This shock wave can have an appreciable effect on the static pressure, extending as it does for a distance equal to a given number of diameters from the nose of the head. One way of overcoming this is to decrease the casing diameter and increase the distance of the static orifices from the nose. Furthermore, a number of orifices may be provided along the length of the pressure head casing so spaced that some will always be in a region of undisturbed airflow.

A long and small-diameter pressure head is an ideal one from an aerodynamic point of view, but it would present certain practical difficulties; its stiffness may not be sufficient to prevent vibration at high speed; and it may also be difficult to accommodate the high-power heater elements required for anti-icing. Thus, in establishing the ultimate relative dimensions of a pressure head, a certain amount of compromise must be accepted.

When a pressure head is at some angle of attack to the airflow, it causes air to flow into the static orifices which creates a pressure above that of the prevailing static pressure, and a corresponding error in static pressure measurement. The pressures developed at varying angles of attack depend on the axial location of the orifices along the casing, their positions around the circumference, their size, and whether the orifices are in the form of holes or slots.

Static Vents

From the foregoing, it would appear that, if all these problems are created by pressure effects only at static orifices, they might as well be separated from the pressure head and positioned elsewhere on the aircraft. This is one solution and is, in fact, put into practice on many types of aircraft by using a pressure head incorporating a pitot tube only, and a static vent in the side of the fuselage. In some light aircraft the vent is simply a hole drilled in the fuselage skin, while for more complex aircraft systems specially contoured metal vent plates are fitted to the skin. A typical pitot tube and static vent are shown in Fig 4.8.

Location of Pressure Heads and Static Vents

For aircraft whose operating ranges are confined to speeds below that of sound some typical locations of pressure heads are ahead of a wing tip, ahead of a vertical stabilizer, or at the side of a fuselage nose section. At speeds above that of sound, a pressure head located ahead of the fuselage nose is, in general, the most desirable location. Independent static vents, when fitted, are always located in the skin of a fuselage, one on each side and interconnected so as to minimize dynamic pressure effects due to yawing or sideslip of the aircraft.

The actual pressure error due to a chosen location is determined for the appropriate aircraft type during the initial flight-handling trials of a prototype, and is finally presented in tabular or graphical form, thus enabling a pilot to apply corrections for various operating conditions. In high-performance aircraft, corrections may be done by a central air data computer (see page 107) which performs the calculations, or in some cases by an aerodynamic method of compensation in which the surfaces in the vicinity either of static vents or of a pressure head are accurately contoured.

ALTERNATE PRESSURE SOURCES

If failure of the primary pitot-static pressure source should occur, for example, complete icing up of a pressure head due to a failed heater circuit, then it is obvious that errors will be introduced in the indications of the instruments and other components dependent on such pressure. As a safeguard against failure, therefore, a standby system may be installed in aircraft employing pressure heads whereby static atmospheric pressure and/or pitot pressure from alternate sources can be selected and connected into the primary system.

The required pressure is selected by means of selector valves connected between the appropriate pressure sources and the flight instruments, and located in the cockpit within easy reach of the flight crew. Figure 4.9 diagrammatically illustrates the method adopted in a system utilizing an alternate static pressure source only. The valves are shown in the normal operating position, i.e. the pressure heads supply pitot and static pressures to the instruments on their respective sides of the aircraft. In the event of failure of static pressure from one or other pressure head the flight instruments are connected to the alternate source of static pressure by manually changing over the position of the relevant selector valve.

The layout shown in Fig 4.10 is based on a system currently in use and is one in which both an alternate source of pitot pressure and static pressure can be selected. Furthermore, it is an example of a system which utilizes the static

64 *Aircraft Instruments*

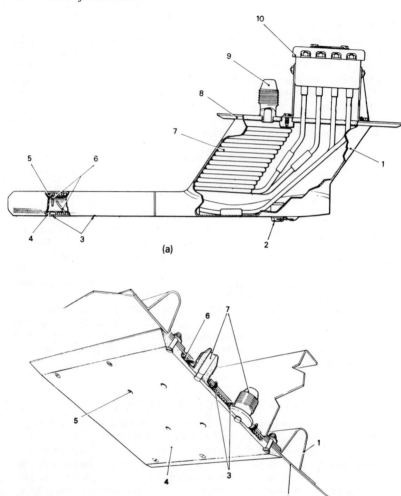

Fig 4.8 Pitot tube and static vent

(a) Pitot tube
1 Mast
2 Drain screw
3 Drain holes
4 Heater element
5 Pressure tube
6 Baffles
7 Heater element
8 Flange plate
9 Pipe adapter
10 Terminal block

(b) Static vent
1 Intercostal
2 Fuselage skin
3 Sealing rings
4 Static vent plate
5 Static vents
6 Thiokol
7 Flanged adaptor boss

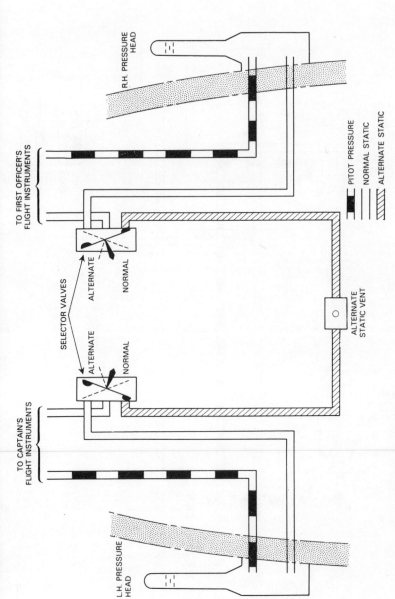

Fig 4.9 Alternate static-pressure system

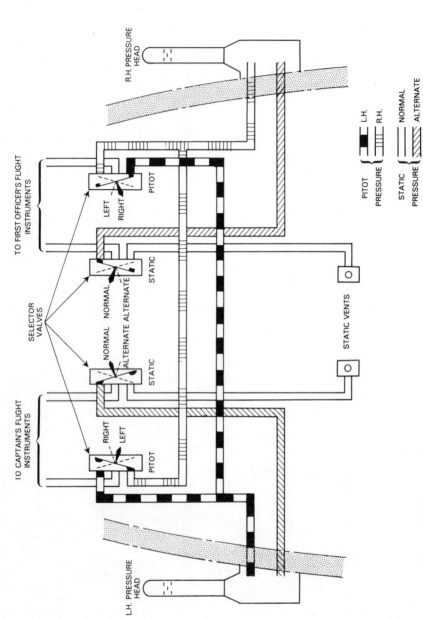

Fig 4.10 Alternate pitot-pressure and static-pressure system

slots of a pressure head as the alternate static pressure source. The valves are shown in the normal position, i.e. the pressure heads supply pitot pressure to the instruments on their respective sides of the aircraft, and the static pressure is supplied from static vent plates. In the event of failure of pitot pressure from one or other pressure head, the position of the relevant selector valve must be manually changed to connect the flight instruments to the opposite pressure head. The alternate static source is selected by means of a valve similar to that employed in the pitot pressure system, and as will be seen from Fig 4.10, it is a straightforward changeover function.

The pressure heads employed in the system just described are of the type illustrated in Fig 4.6, reference to which shows that two sets of static slots (front and rear) are connected to separate pipes at the mounting base. In addition to being connected to their respective selector valves, the pressure heads are also coupled to each other by a cross-connection of the static slots and pipes; thus, the front slots are connected to the rear slots on opposite heads. This balances out any pressure differences which might be caused by the location of the static slots along the fore-and-aft axis of the pressure heads.

DRAINS

In order for a pitot-static system to operate effectively under all flight conditions, provision must also be made for the elimination of water that may enter the system as a result of condensation, rain, snow, etc., thus reducing the probability of "slugs" of water blocking the lines. Such provision takes the form of drain holes in pressure heads, drain traps, and drain valves in the system pipelines. Drain holes are drilled in pressure-head pitot tubes and casings (Fig 4.5), and are of such a diameter that they do not introduce errors in instrument indications.

The method of draining the pipelines of a pitot-static system varies between aircraft types because one aircraft manufacturer may fabricate his own design of drain trap and valve, while another may use a prefabricated component from a specialized manufacturer. Some typical drain traps and valves in current use are illustrated in Fig 4.11.

Drain traps are designed to have a capacity sufficient to allow for the accumulation of the maximum amount of water that could enter the system between servicing periods.

The drain valves are of the self-closing type so that they cannot be inadvertently left in the open position after drainage of accumulated water.

PIPELINES

Pitot and static pressures are transmitted through seamless and corrosion-resistant metal (light alloy and/or tungum) pipelines and flexible pipes, the latter being used at the instrument panel to allow for movement on the antivibration mountings.

The chosen diameter of pipelines is related to the distance from the pressure sources to the instruments (the longer the lines, the larger the diameter) in order to eliminate pressure drop and time-lag factors. There is, however, a minimum acceptable limit to the internal diameter, namely $\frac{1}{4}$ in. A smaller

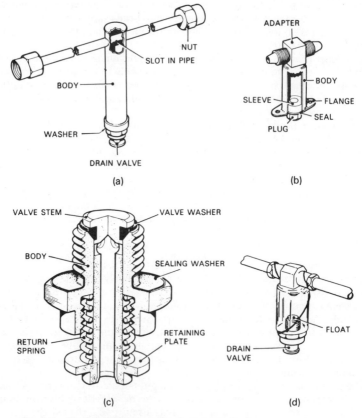

Fig 4.11 Pitot-static system water-drainage methods
(a) Water trap and drain valve   (c) Drain valve construction
(b) Water trap and drain plug   (d) Transparent water trap and drain valve

internal diameter would present the hazard of a blockage due to the probability of a "slug" of water developing in such a way as to span the diameter.

Identification of the pipelines is given by conventional colour-coded tapes spaced at frequent intervals along the lines. In addition to this method, the lines in some aircraft may also be of different diameters.

MEASUREMENT OF ALTITUDE

The Earth's Atmosphere and Characteristics

The earth's atmosphere, as most readers no doubt already know, is the surrounding envelope of air, which is a mixture of a number of gases the chief of which are nitrogen and oxygen. This gaseous envelope is conventionally divided into several concentric layers extending from the earth's surface, each with its own distinctive features.

The lowest layer, the one in which we live and in which conventional types of aircraft are mostly flown, is termed the *troposphere* and extends to a height of about 28,000 ft at the equator. This height boundary is termed the *tropopause*.

Above the tropopause, we have the layer termed the *stratosphere* extending to the *stratopause* at an average height between 60 and 70 miles.

At greater heights the remaining atmosphere is divided into further layers or regions which from the stratopause upwards are termed the *ozonosphere*, *ionosphere* and *exosphere*.

Throughout all these layers the atmosphere undergoes a gradual transition from its characteristics at sea-level to those at the fringes of the exosphere where it merges with the completely airless outer space.

Atmospheric Pressure

The atmosphere is held in contact with the earth's surface by the force of gravity, which produces a pressure within the atmosphere. Gravitational effects decrease with increasing distances from the earth's centre, so that atmospheric pressure decreases steadily with altitude. The units used in expressing atmospheric pressure are: pounds per square inch, inches of mercury and millibars.

The standard sea-level pressure is 14·7 lb/in^2 and is equal to 29·921 in Hg or 1013·25 mb. Thus, a column of air one square inch in section, extending from the earth's surface to the extremities of the atmosphere, weighs 14·7 lb and so exerts this pressure on one square inch of the earth's surface. Suppose now that this same column of air is extended from points 5,000, 10,000 and 15,000 ft above sea-level; the weight will have decreased and the pressures exerted at these levels are found to be 12·2, 10·1 and 8·3 lb/in^2 respectively. At the tropopause, the pressure falls to about a quarter of its sea-level value.

The steady fall in atmospheric pressure as altitude increases has a dominating effect on the density of the air, which changes in direct proportion to changes of pressure.

Atmospheric Temperature

Another important factor affecting the atmosphere is its temperature characteristic. The air in contact with the earth is heated by conduction and radiation, and as a result its density decreases and the air starts rising. In rising, the pressure drop allows the air to expand, and the expansion in turn causes a fall in temperature. Under standard sea-level conditions the temperature is 15°C, and falls steadily with increasing altitude up to the tropopause, at which it remains constant at −56·5°C. The rate at which it falls is termed the *lapse rate*.* In the stratosphere the temperature at first remains constant at −56·5°C, then it increases again to a maximum at a height of about 40 miles, after which it starts to decrease reaching freezing point at about 50 miles. From this altitude there is yet a further increase reaching a maximum of about 2200°C at about 150 miles.

Standard Atmosphere

In order to obtain indications of altitude, airspeed and rate of altitude change it is of course necessary to know the relationship between the pressure, temperature and density variables, and altitude. Now, for such indications to be

* From the Latin *lapsus*, slip

70 Aircraft Instruments

presented with absolute accuracy, direct measurements of the three variables would have to be taken at all altitudes and fed into the appropriate instruments as correction factors. Such measurements, while not impossible, would, however, demand some rather complicated instrument mechanisms. It has therefore always been the practice to base all measurements and calculations related to aeronautics on what is termed a *standard atmosphere*, or one in which the values of pressure, temperature and density at the different altitudes are assumed to be constant. These assumptions have in turn been based on established meteorological and physical observations, theories and measurements, and so the standard atmosphere is accepted internationally. As far as altimeters, airspeed indicators and vertical-speed indicators are concerned, the inclusion of the assumed values of the relevant variables in the laws of calibration permits the use of simple mechanisms operating solely on pressure changes.

The assumptions are as follows: (i) the atmospheric pressure at mean sea-level is equal to 1013·25mb or 29·921 in Hg; (ii) the temperature at mean sea-level is 15°C (59°F); (iii) the air temperature decreases by 1·98°C for every 1,000 ft increase in altitude (this is the lapse rate referred to above) from 15°C at mean sea-level to $-56·5°C$ (69·7°F) at 36,089 ft. Above this altitude the temperature is assumed to remain constant at $-56·5°C$.

It is from the above sea-level values that all other corresponding values have been calculated and presented as the standard atmosphere. These calculated values were originally defined as ICAN (International Commission for Aerial Navigation) conditions, but in 1952 a more detailed specification was finally established by the International Civil Aviation Organization, and so the *ICAO Standard Atmosphere* is the one now accepted.

Measurement of Atmospheric Pressure

There are two principal methods of measuring the pressure of the atmosphere, both of which are closely associated with pitot-static flight instruments. They are (i) balancing of pressure against the weight of a column of liquid, and (ii) magnifying the deflection of an elastic sensing element produced by the pressure acting on it. The instruments employing these two methods are known respectively as the mercury barometer and the aneroid barometer.

Mercury Barometer

As shown in Fig 4.12 (*a*), a *mercury barometer* consists essentially of a glass tube sealed at one end and mounted vertically in a bowl or cistern of mercury so that the open end of the tube is submerged below the surface of the mercury. Let us imagine for the moment that the upper end of the tube is open as shown at (*b*) and that an absolute pressure p_1 is applied to the mercury in the cistern, and an absolute pressure p_2, to the column of mercury within the tube. If p_1 is greater than p_2 then obviously mercury will be forced down in the cistern and will rise in the tube until a balance between the two pressures is achieved. This balance is given by

$$p_1 = p_2 + H\rho \tag{1}$$

where H is the difference in levels between the mercury in the cistern and the tube, and ρ is the density of mercury.

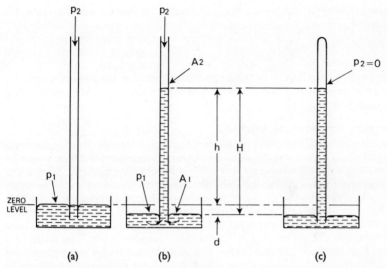

Fig 4.12 Principle of mercury barometer

But, as may be seen from the diagram, $H = h + d$, where h is the distance the mercury has risen in the tube from the zero level, and d is the corresponding distance the mercury has fallen in the cistern from the zero level.

The same quantity of mercury has left the cistern and entered the tube and is such that, if the sectional areas of the cistern and tube are designated A_1 and A_2, respectively, then

$$A_1 d = A_2 h \qquad (2)$$

so that

$$d = \frac{A_2}{A_1} h \qquad (3)$$

Therefore

$$H = h + h \frac{A_2}{A_1} \qquad (4)$$

Replacing H in eqn (1) by this expression,

$$p_1 = p_2 + h \left(1 + \frac{A_2}{A_1}\right) \rho \qquad (5)$$

In the practical case the upper end of the tube is sealed and a vacuum*

---

* Known as the Torricellian vacuum, after Torricelli, the Italian inventor of the barometer.

exists in the space above the mercury column, so that p_2 is zero and eqn (5) becomes

$$p_1 = h \left(1 + \frac{A_2}{A_1}\right) \rho \qquad (6)$$

which forms the basic barometer equation, in which h, the height the mercury rises in the tube, is a measure of the absolute pressure applied to the mercury in the cistern.

Aneroid Barometer and Altimeter

In the general development of instruments for atmospheric pressure measurement, certain practical applications demanded instruments which would be portable and able to operate in various attitudes. For example, for weather observations at sea it was apparent that a mercury barometer would be rather fragile, and that under pitching and rolling conditions the erratic movement of the mercury column would make observations difficult. A more robust instrument was therefore needed and it had to be one which required no liquid whatsoever. Thus the *aneroid* barometer* came into being. The instrument has been developed to quite a high standard of precision for many specialized applications, but the simplest version of it, and one more familiar perhaps to the reader, is the household barometer.

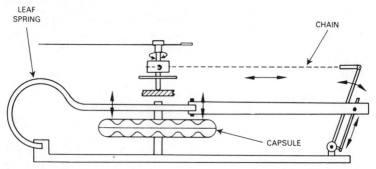

Fig 4.13   Aneroid barometer

The pressure-sensing element of the instrument (see Fig. 4.13) is an evacuated metal capsule. Since there is approximately zero pressure inside the capsule, and assuming the instrument to be at sea-level, approximately 14.7 lb/in^2 on the outside, the capsule will tend to collapse. This, however, is prevented by a strong leaf spring fitted so that one side is attached to the top of the capsule and the other side to the instrument baseplate. The spring always tends to open outwards, and a state of equilibrium is obtained when the pressure is balanced by the spring tension.

If now the atmospheric pressure decreases, the force tending to collapse the capsule is decreased but the spring tension remains the same and consequently is able to open out the capsule a little further than before. If there is an increase in pressure, the action is reversed, the pressure now collapsing the capsule against the tension of the spring until equilibrium is attained.

* From the Greek *aneros*, not wet.

The resulting expansion and contraction of the capsule, which is extremely small, is transformed into rotary motion of the pointer by means of a magnifying lever system and a very finely-linked chain.

From the foregoing description, we can appreciate that when it first became necessary to measure the height of an aeroplane above the ground, the aneroid barometer with a change of scale markings formed a ready-made altimeter. Present-day altimeters are, of course, much more sophisticated, but the aneroid barometer principle still applies. The mechanism of a typical sensitive altimeter is shown in Fig 4.14. The pressure-sensing element is made up of three aneroid capsules stacked together to increase the sensitivity of the instrument. Deflections of the capsules are transmitted to a sector gear via a link and rocking shaft assembly. The sector gear meshes with a magnifying gear mechanism which drives a handstaff carrying a long pointer the function of which is to indicate hundreds of feet. A pinion is also mounted on the handstaff, and this drives a second gear mechanism carrying second and third pointers which indicate thousands and tens of thousands of feet respectively. In this particular instrument a disc is also attached to the third pointer gear and moves with it. One side of the disc is painted white, and above 10,000 ft this becomes visible through a semicircular slot cut in the main dial. Thus, the pointer movement is "traced" out to eliminate ambiguity of readings above 10,000 ft.

The pressure-sensing element is compensated for changes in ambient temperature by a bimetal U-shaped bracket, the open ends of which are connected to the top capsule by means of push rods. Temperature changes which affect the flexibility of the capsules similarly affect the bracket, causing slight movement of its side arms. The push rods then transmit the movement to the capsules to counteract their tendency to introduce errors as a result of temperature changes.

A barometric pressure-setting mechanism, the purpose of which is described on page 77, is mounted in front of the main mechanism. It consists of a counter geared to the shaft of a setting knob. The shaft also carries a pinion which meshes with a gear around the periphery of the main mechanism casting. When the knob is rotated to set the required barometric pressure, the main mechanism is also rotated, and the pointers are set to the corresponding altitude change. The position of the capsules under the influence of the atmospheric pressure prevailing at the time of setting remains undisturbed. A spring-loaded balance weight is linked to the rocking shaft to maintain the balance of the main mechanism regardless of its attitude.

Altimeter Dial Presentations

The presentation of altitude information has undergone many changes in recent years principally as a result of altimeter misreading being the proven or suspected cause of a number of fatal accidents. In consequence, several methods are to be found in altimeters currently in use, the most notable of which are the triple-pointer, single-pointer and digital counter, and single-pointer and drum presentations. The triple-pointer method is the oldest of presentations and is the one which really made it necessary to introduce changes. This method is used in the altimeter shown in Fig 4.14: the susceptibility of its predecessor to misreading of 1,000 ft and 10,000 ft, has been overcome to a large extent by giving the pointers a more distinctive shape, and by

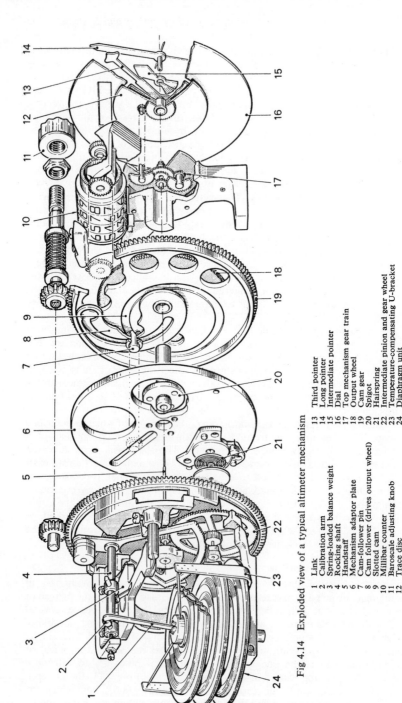

Fig 4.14 Exploded view of a typical altimeter mechanism

1 Link
2 Calibration arm
3 Spring-loaded balance weight
4 Rocking shaft
5 Handstaff
6 Mechanism adaptor plate
7 Cam-follower pin
8 Cam follower (drives output wheel)
9 Slotted cam
10 Millibar counter
11 Baroscale adjusting knob
12 Trace disc
13 Third pointer
14 Long pointer
15 Intermediate pointer
16 Dial
17 Top mechanism gear train
18 Output wheel
19 Cam gear
20 Spigot
21 Hairspring
22 Intermediate pinion and gear wheel
23 Temperature-compensating U-bracket
24 Diaphragm unit

incorporating the trace disc referred to earlier. In addition, some versions incorporate a yellow and black striped disc which serves as a low-altitude warning device by coming into view at altitudes below 16,000 ft.

The counter/pointer (Fig. 4.15(c)), and in some cases the drum/pointer, presentations are used in servo altimeters and altimeters forming part of air data systems (see pages 79 and 103).

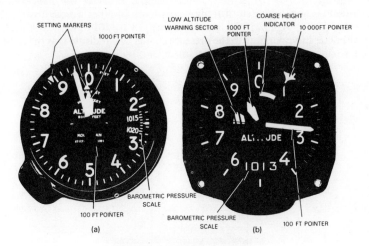

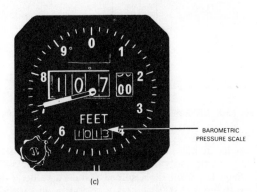

Fig 4.15 Altimeter dial presentations
        (a) Triple-pointer (10,000 ft pointer behind 1,000 ft pointer in this view)
        (b) Modified triple pointer
        (c) Counter/pointer

Errors Due to Changes in Atmospheric Pressure and Temperature

As we already know, the basis for the calibration of altimeters is the standard atmosphere. When the atmosphere conforms to standard values, an altimeter will read what is termed *pressure altitude*. In a non-standard atmosphere, an altimeter is in error and reads what is termed *indicated altitude*.

76 Aircraft Instruments

We may consider these errors by taking the case of a simple altimeter situated at various levels. In standard conditions, and at a sea-level airfield, an altimeter would respond to a pressure of 1013·25 mb (29·92 in Hg) and indicate the pressure altitude of zero feet. Similarly, at an airfield level of 1,000 ft, it would respond to a standard pressure of 977·4 mb (28·86 in Hg) and indicate a pressure altitude of 1,000 ft. Assuming that at the sea-level airfield the pressure falls to 1012·2 mb (29·89 in Hg), the altimeter will indicate that the airfield is approximately 30 ft above sea-level; in other words, it will be in error by +30 ft. Again, if the pressure increases to 1014·2 mb (29·95 in Hg), the altimeter in responding to the pressure change will indicate that the airfield is approximately 30 ft below sea-level; an error of −30 ft.

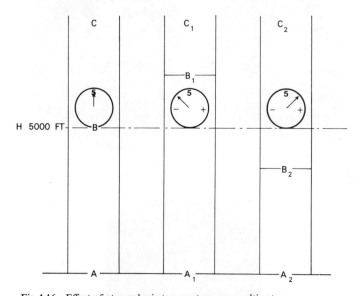

Fig 4.16 Effect of atmospheric temperature on an altimeter

In a similar manner, errors would be introduced in the readings of such an altimeter in flight and whenever the atmospheric pressure at any particular altitude departed from the assumed standard value. For example, when an aeroplane flying at 5,000 ft enters a region in which the pressure has fallen from the standard value of 842·98 mb to, say, 837 mb, the altimeter will indicate an altitude of approximately 5,190 ft.

The standard atmosphere also assumes certain temperature values at all altitudes and consequently non-standard values can also cause errors in altimeter readings. Variations in temperature cause differences of air density and therefore differences in weight and pressure of the air. This may be seen from the three columns shown in Fig 4.16. At point A the altimeter measures the weight of the column of air above it, of height AC to the top of the atmospheric belt. At point H which is, say, at an altitude of 5,000 ft above A, the weight or pressure on the altimeter is less by the weight of the part AB of the column below H. If the temperature of the air in the part AB rises, the column will expand to A_1B_1, and at H the pressure on the altimeter is now less by the weight of A_1H. But the weight of A_1B_1 is still the same as that of AB, and therefore the weight of A_1H must be less than that of AH, and so the altimeter in rising from A_1 to 5,000 ft will register a smaller reduction of pressure than when it rose from point A to 5,000 ft. In other words, it will read less than 5,000 ft. Similarly, when the temperature of the air between points A and H decreases, the part AB of the column shrinks to A_2B_2 and the change of pressure on the altimeter in rising from A_2 to 5,000 ft will be not only the weight of A_2B_2 (which equals AB) but also the weight of B_2H. The altimeter

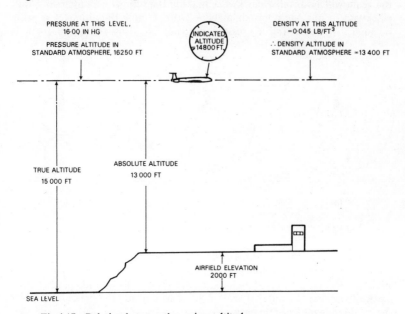

Fig 4.17   Relation between the various altitudes

will thus measure a greater pressure drop and will indicate an altitude greater than 5,000 ft. The relationship between the various altitudes associated with aircraft flight operations are presented graphically in Fig 4.17.

It will be apparent from the foregoing that, although the simple form of altimeter performs its basic task of measuring changes in atmospheric pressure accurately enough, the corresponding altitude indications are of little value unless they are corrected to standard pressure datums. In order, therefore, to compensate for altitude errors due to atmospheric pressure changes, altimeters are provided with a manually operated adjustment device which allows the pointers to be set to zero height for any prevailing ground pressure so that the indications in flight will still be heights in the standard atmosphere above the ground.

The adjustment device consists basically of a scale or counter calibrated in millibars or inches of mercury which is interconnected between a setting knob

Aircraft Instruments

and the altitude indicating mechanism in such a way that the correct pressure/height relationship is obtained. The underlying principle may be understood by referring to Fig 4.18, which shows a scale-type device in very simple form. The scale is mounted on a gear which meshes with a pinion on the end of the control knob shaft and also with the pointer gearing. A differential gear mechanism (not shown) allows the pointer to be rotated without disturbing the setting of the capsules.

In Fig 4.18 (*a*) the altimeter is assumed to be subjected to standard conditions; thus the millibar scale when set to 1013 mb positions the pointer at the 0 ft graduation. If the millibar setting is changed from 1013 to 1003 as at (*b*), the scale will be rotated clockwise, making the altimeter pointer turn anticlockwise to indicate approximately −270 ft. If now the altimeter is raised through 270 ft as at (*c*), a pressure decrease of 10 mb (1013−1003) will be

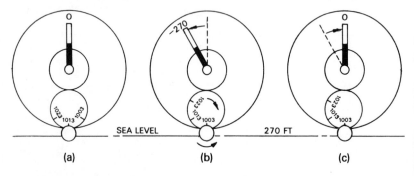

Fig 4.18 Principle of barometric pressure adjustment

measured by the capsule and the pointer will return to zero. Thus whatever pressure is set on the millibar scale, the altimeter will indicate zero when subjected to that pressure. Similarly, any setting of the altitude pointer automatically adjusts the millibar scale reading to indicate the pressure at which the height indicated will be zero.

In practice, the method of setting is a little more complicated because the relationship between pressure and altitude is non-linear. Therefore, in order to obtain linear characteristics of the required pressure and altitude readings, an accurately profiled cam correction device (see Fig 4.14) is connected between the pressure counter and main mechanism.

"Q" Code for Altimeter Setting

The setting of altimeters to the barometric pressures prevailing at various flight levels and airfields is part of flight operating techniques, and is essential for maintaining adequate altitude separation between aircraft, and terrain clearance during take-off and landing. In order to make the settings a pilot is dependent on observed meteorological data which is requested and transmitted from ground control centres. The requests and transmissions are adopted universally and form part of the ICAO "Q" code of communication. Three code letter groups are normally used in connection with altimeter settings, and are defined as follows:

QFE Setting the pressure prevailing at an airfield to make the altimeter read zero on landing and take-off.
QNE Setting the standard sea-level pressure of 1013·25 mb (29·92 in Hg) to make the altimeter read the airfield elevation.
QNH Setting the pressure scale to make the altimeter read airfield height above sea-level on landing and take-off.

Servo Altimeter

The mechanism of a typical servo altimeter is shown schematically in Fig 4.19, from which it will be noted that the pressure-sensing capsule element is coupled to an electrical pick-off assembly instead of a mechanical linkage system as in conventional altimeters. The inductive type of pick-off consists of a pivoted laminated I-bar coupled to the capsules and positioned at a very small distance from the limbs of a laminated E-bar pivoted on a cam follower. A coil is wound around the centre limb of the E-bar and is supplied with alternating current, while around the outer limbs coils are wound and connected in series to supply an output signal to an external amplifier unit. Thus, the pick-off is a special form of transformer, the centre-limb coil being the primary winding and the outer-limb coils the secondary winding.

A two-phase drag-cup type of motor is coupled by a gear train to the pointer and counter assembly, and also to a differential gear which drives a cam. The cam bears against a cam follower so that as the cam position is changed the E-bar position relative to the I-bar is altered. The reference phase of the motor is supplied with a constant alternating voltage from the main source, and the control phase is connected to the amplifier output channel.

Setting of barometric pressure is done by means of a setting knob geared to a digital counter, and through a special rod and lever mechanism, to the differential gear and cam. Thus, rotation of the setting knob can also alter the relative positions of the E-bar and I-bar.

A double-contact switch is provided within the case and is connected in the power supply circuit to interrupt the latter should the servomotor overrun. A solenoid-operated warning flag comes into view whenever an overrun occurs and for any other condition causing an interruption of the power supply.

When the aircraft altitude changes the capsules respond to the changes in static pressure in the conventional manner. The displacement of the capsules is transmitted to the I-bar, changing its angular position with respect to the E-bar and therefore changing the air-gaps at the outer ends. This results in an increase of magnetic flux in one outer limb of the E-bar and a decrease in the other. Thus, the voltage induced in one of the secondary coils will increase, while in the other it decreases. An output signal is therefore produced at the secondary coil terminals, which will be either in phase or out of phase with the primary-coil voltage, depending on the direction of I-bar displacement. The magnitude of the signal will be governed by the magnitude of the deflection.

The signal is fed to the amplifier, in which it is amplified and phase detected, and then supplied to the servomotor control winding. The motor rotates and drives the pointer and height-counter mechanism in the direction appropriate to the altitude change. At the same time, the servomotor gear train rotates a worm-gear shaft and the differential gear which is meshed with it. The cam and cam follower are therefore rotated to position the E-bar in a direction

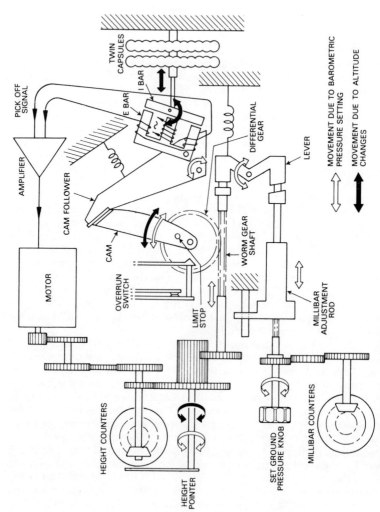

Fig 4.19 Typical servo altimeter mechanism

which will cause the magnetic fluxes in the cores, and the secondary-coil voltages, to start balancing each other. When the E-bar reaches the null position, i.e. when the aircraft levels off at a required altitude, no further signals are fed to the amplifier, the servomotor ceases to rotate, and the pointer and counters indicate the new altitude.

When the barometric-pressure setting knob is rotated, the pressure counters are turned and the lever of the setting mechanism moves the worm-gear shaft laterally. This movement of the shaft rotates the differential gear, cam and cam follower, causing relative displacement between the E-bar and I-bar. An error signal is therefore produced which, after amplification and phase detection, drives the servomotor and gear mechanisms in a sequence similar to that resulting from a normal altitude change. When the null position of the E-bar is reached, however, the pointer and counters will indicate aircraft altitude with respect to the barometric pressure adjustment.

Cabin Altimeters

In pressurized-cabin aircraft, it is important that the pilot and crew have an indication that the cabin altitude corresponding to the maximum differential pressure conditions is being maintained. To meet this requirement, simple altimeters, calibrated to the same pressure/altitude law as normal altimeters, are provided on the main instrument panel, or on the pressurization system control panel, their measuring elements responding directly to the prevailing cabin air pressure.

Fig 4.20 shows the mechanism and dial presentation of a typical instrument. The case is sealed except for a vent at the rear to admit the cabin air. As the cabin air pressure increases, the aneroid capsule stack is compressed, causing the rocking shaft and gear mechanism to rotate the pointer to indicate the selected cabin pressure altitude.

Altitude Switches and Warning Systems

In certain aircraft systems the control and operating conditions are related to one specific altitude, and it is necessary to provide a visual indication, or an aural warning that such altitude has been attained. For example, in cabin-pressurization systems, the necessity arises for an indication of a possible increase of cabin altitude above the desired level while the aircraft is at its normal operating altitude. Also it may be necessary to warn a pilot that his aircraft is passing through critical altitudes and so assist him in overcoming the danger of misinterpreting altimeter indications.

To cater for requirements of this nature, an aneroid-capsule measuring element similar to that used in altimeters is adapted as an altitude switching and warning device, and constructed to make and break the circuit in which it is connected, at the specified altitude.

An example of an altitude switch in practical use as an indication system is shown in Fig 4.21. The pressure-sensing element is made up of a capsule soldered into a brass base, the capsule being acted upon by the air entering the unsealed plastic case. A dome-headed screw is positioned on the centre-piece of the capsule and is pre-adjusted to the required altitude setting. The switch assembly comprises two contacts which are adjusted to a fixed gap setting, and a movable leaf-spring contact. The upper fixed-contact gap setting

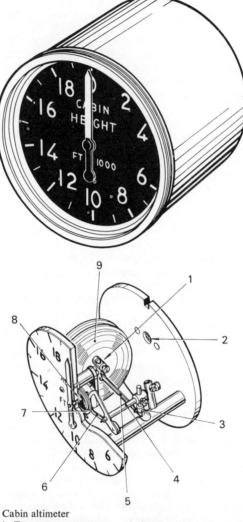

Fig 4.20 Cabin altimeter
1 Temperature compensator
2 Filtered vent
3 Calibrating arm
4 Rocking shaft assembly
5 Link
6 Sector
7 Hairspring
8 Handstaff
9 Diaphragm assembly

allows for the difference between the pressures at which make and break takes place, while the lower contact-gap setting determines the amount of force required to trip the movable contact.

When the pre-determined cabin altitude has been reached, the deflection of the capsule and dome-headed screw causes the spring contact, which is in constant tension, to move over with a snap action to a fixed contact, thus making or breaking a circuit depending on the electrical connection arrangements.

Pitot-static Instruments and Systems 83

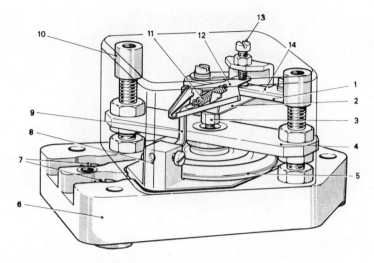

Fig 4.21 Typical altitude switch unit

1	Insulating plate	8	Mounting plate
2	Switch cantilever	9	Spring post
3	Dome-headed adjusting screw	10	Threaded pillar
4	Overload bridge	11	Tension spring
5	Capsule	12	Switch tongue
6	Base plate	13	Differential adjusting screw
7	Terminal screws	14	Contact plate assembly

An example of altitude switch designed to form part of a true warning system is illustrated in Fig 4.22. The complete system consists of the switch unit, a flasher unit and two white lights, and a pushbutton switch located on the main instrument panel. The pressure-sensing element is an aneroid capsule subjected to static pressure from the aircraft's pitot-static system. The switch assembly is a double-contact unit, one contact being on the capsule and the other at one end of an adjustable spring-loaded plunger. A scale graduated in thousands of feet is provided to indicate the altitude for which the plunger and contacts have been set.

In addition to the foregoing assemblies, two miniature relays designated as A and B in Fig 4.23 are mounted within the casing. The flasher unit is a transistorized assembly and is incorporated within an insulated block located in the switch unit casing.

During take-off and the ensuing climb, the capsule contacts are open, both relays are de-energized and the lamps are extinguished. As the aircraft passes through the altitude preset in the system, the capsule under the influence of decreasing static pressure will have expanded sufficiently to close its contacts and to energize relay A directly from the direct-current supply. A circuit is thus completed through contacts 1 and 2 of the de-energized relay B, contacts 1 and 3 of relay A to the flasher unit and ground. The flasher unit commences to operate and supplies current intermittently to both warning lights, which start flashing to indicate to the pilots that the aircraft is now above the preset altitude.

Fig 4.22 Altitude switch for use in a warning system

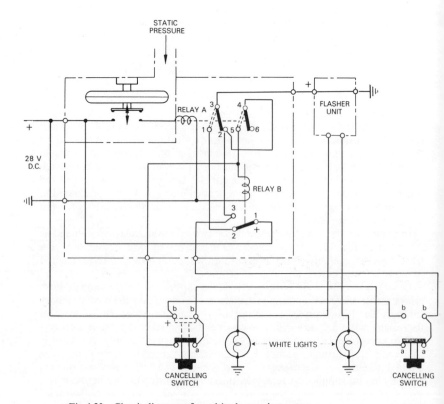

Fig 4.23 Circuit diagram of an altitude warning system

Once above this altitude, continued operation of the flasher unit and lights is unnecessary, and so a cancelling facility is provided by means of the pushbutton switches. Assuming that the captain presses his cancelling switch then a positive current will flow across contacts b of the switch to contacts a of the co-pilot's switch. From these contacts the current flows to contact 3 of relay B, across contacts 4 and 5 of the energized relay A, and finally to the coil of relay B. On being energized, this relay changes over its contacts from 1 and 2 to 1 and 3 and so interrupts the circuit to the flasher unit and lights. In a similar manner cancelling can be effected by the other pushbutton switch.

For obvious reasons, it should be unnecessary for a pilot to have to continually hold the switch button depressed; an automatic "hold-in" facility is therefore incorporated in the circuit and operates in the following way. At the instant when the contacts of relay B change over, the current at contact 1 which was operating the flasher unit now flows from contact 3 through the same circuit as that set up by the push switch, namely the coil circuit of relay B. Thus, at that instant the relay will be energized by a current flowing over its own closed contacts. In this condition the pushbutton may be released, because the current to the coil of relay B also flows to contacts a of the switch, and tracing this through on Fig 4.23, we see that it holds the relay energized.

During descent, the capsule contracts in response to the increasing static pressure until at the preset altitude the contacts open and relay A de-energizes. The opening of contacts 4 and 5 now interrupts the circuit to the coil of relay B, but due to the "hold-in" circuit it will remain energized. A positive current is now fed from contact 3 of this relay to the flasher unit via contacts 2 and 3 of relay A, so the lights again start flashing to indicate that the aircraft is now below the preset altitude.

If, in this operating condition, the captain presses his cancelling switch, the "hold-in" circuit across switch contacts a is interrupted causing relay B to de-energize. This, in turn, interrupts the circuit to the flasher unit and lights, and returns the whole system to its starting condition.

AIRSPEED INDICATORS

Airspeed indicators are in effect very sensitive pressure gauges measuring the difference between the pitot and static pressures detected by the pressure head, in terms of the $\frac{1}{2}\rho V^2$ formula given on page 59.

Originally this formula served as the basis for the calibration of indicators, but as airspeeds began to increase a noticeable error began to creep into the indications. The reason for this was that, as the high-velocity air was brought to "stagnation" at the pitot tube, it was compressed and its density was increased. The formula was therefore modified to take into account additional factors in order to minimize the "compressibility error." Thus, present-day airspeed indicators are calibrated to the law

$$p = \tfrac{1}{2}\rho V^2 \left(1 + \tfrac{1}{4}\frac{V^2}{a_0^2}\right)$$

where p = Pressure difference (mm H$_2$O)
 ρ = Density of air at sea-level
 V = Speed of aircraft (m.p.h. or knots)
 a_0 = Speed of sound at sea-level (m.p.h.)

86 *Aircraft Instruments*

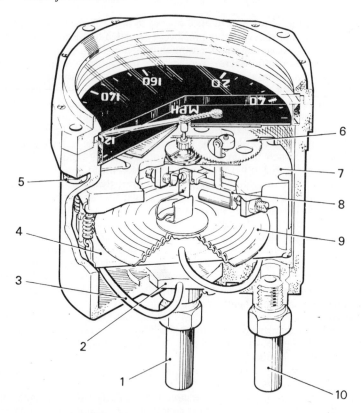

Fig 4.24 Simple type of airspeed indicator

1 Pitot connector
2 Pitot union
3 Capsule capillary
4 Capsule plate
5 Locknut
6 Pointer movement assembly
7 Frame casting
8 Rocking shaft assembly
9 Capsule assembly
10 Static connector

The numerical values to be inserted in the formula depend on whether V is expressed in m.p.h. or knots*:

$$p = 0.012504\, V^2 (1 + 0.43 V^2 \times 10^{-6}) \text{ with } V \text{ in m.p.h.}$$

$$= 0.016580 V^2 (1 + 0.57 V^2 \times 10^{-6}) \text{ with } V \text{ in knots}$$

Both formulae take into account the relative densities of air and water.

The mechanism of a typical simple-type airspeed indicator is illustrated in Fig 4.24. The pressure-sensing element is a metal capsule the interior of which

---

* The *knot* is the unit commonly used for expressing the speeds of aircraft:

1 knot = 1 nautical mile per hour
= 1·15 m.p.h.
= 6,080 ft/h

is connected to the pitot pressure connector via a short length of capillary tube which damps out pressure surges. Static pressure is exerted on the exterior of the capsule and is fed into the instrument case via the second connector. Except for this connector the case is sealed.

Displacements of the capsule in accordance with what is called the "square-law" are transmitted via a magnifying lever system, gearing, and a square-law compensating device to the pointer, which moves over a scale calibrated in knots. The purpose of the compensator is to provide a linear scale, and the principles of three commonly used methods are described on page 89. Temperature compensation is achieved by a bimetallic strip arranged to vary the magnification of the lever system in opposition to the effects of temperature on system and capsule sensitivity.

Square-law Compensation

Since airspeed indicators measure a differential pressure which varies with the square of the airspeed, it follows that, if the deflections of the capsules responded linearly to the pressure, the response characteristic in relation to speed would be similar to that shown in Fig 4.25 (a). If also the capsule were

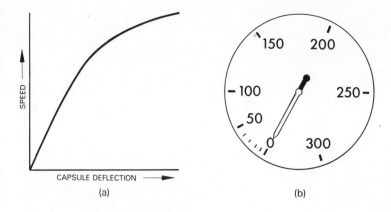

Fig 4.25 Square-law characteristics
(a) Effect of linear deflection/pressure response
(b) Effect of direct magnification

coupled to the pointer mechanism so that its deflections were directly magnified, the instrument scale would be of the type indicated at (b).

The non-linearity of such a scale makes it difficult to read accurately, particularly at the low end of the speed range; furthermore, the scale length for a wide speed range would be too great to accommodate conveniently in the standard dial sizes.

Therefore, to obtain the desired linearity a method of controlling either the capsule characteristic, or the dimensioning of the coupling element conveying capsule deflections to the pointer, is necessary. Of the two methods the latter

88 *Aircraft Instruments*

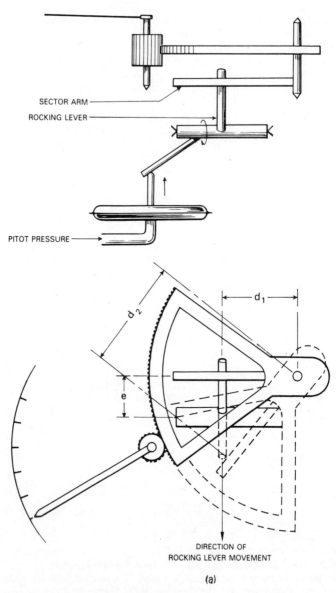

Fig 4.26 Methods of square-law compensation
(a) Rocking-lever/sector-arm mechanism   (b) "Banana" slot

Pitot-static Instruments and Systems   89

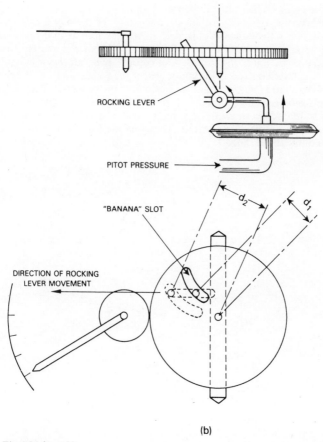

Fig 4.26 (contd.)

is the more practical because means of adjustment can be incorporated to overcome the effects of capsule "drift" plus other mechanical irregularities as determined during calibration.

There are two versions of this method in common use, the principle being the same in both cases, i.e. the length of a lever is altered as progressive deflections of the capsule take place, causing the mechanism and pointer movement to be increased for small deflections and decreased for large deflections. In other words, it is a principle of variable magnification.

The lever length referred to is the distance between the axis of the main gear of the mechanism and the point of contact between its rocking lever and the gear; this distance is indicated as d_1 and d_2 in Figs 4.26 (a) and (b).

The method shown at (a) may be considered as a basic one and serves as a very useful illustration of the operating principle. At the starting position of the mechanism, the rocking lever and the sector arm are in contact with each other at an angle preset for the ranging of the instrument; therefore setting the distance d_1. When pitot pressure is applied to the capsule, the latter is deflected causing the rocking shaft to rotate, and the rocking lever to move in a straight line in the direction indicated. As may be seen from the diagram, the rocking lever pushes the sector arm round and distance d_1 starts increasing. The initial deflection of the capsule is of course small, but this is magnified by the rocking lever contacting the sector arm at the distance e from the centre of the rocking shaft. Thus, a magnified movement of the sector arm is obtained, and the sector gear and pinion in turn provide further magnification to the pointer so that it will travel through a large distance for the small pressure/deflection characteristic of the capsule. Therefore the scale is "opened up" at the low end of the speed range.

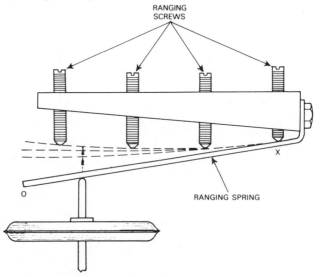

Fig 4.27 Spring-type of square-law compensator

OX = effective spring length diminishing as spring makes contact with screws

Assuming that the pitot pressure continues to increase, the capsule will deflect at an increasing rate and the rocking lever movement will follow the square-law deflections, but as its point of contact moves through the distance d_2, its force acts further and further along the sector arm and decreases the rotational movement of the latter and also the magnification of the pointer mechanism.

Therefore, a "closing up" of the scale is obtained for increasing airspeed, the initial settings of the whole arrangement being such that, in following what may be termed a "square-root law," a linear scale is produced throughout the speed range for which the instrument is calibrated.

In the mechanism shown in Fig 4.26 (b), the sector gear and arm are replaced by a large circular gear and an integrally cut radial slot which, for obvious reasons, is termed a "banana slot." The rocking lever engages with this slot so that as the lever moves it rotates the gear and changes the slot position, which decreases the magnification in exactly the same manner as in method (a).

A third type of square-law compensating device is shown in Fig 4.27. It consists of a special ranging or "tuning" spring which bears against the capsule and applies a controlled retarding force to capsule expansion. The retarding force is governed by sets of ranging screws which are pre-adjusted to contact the spring at appropriate points as it is lifted by the expanding capsule. As the speed and differential pressure increase, the spring rate increases and its effective length is shortened; thus linearity is obtained directly at the capsule and not at the magnifying lever system as in Fig 4.26.

MACHMETERS

With the advent of the gas turbine, the propulsive power available made it possible for greater flying speeds to be attained, but at the same time certain limiting factors related to the strength of an airframe structure, and the forces acting on it, were soon apparent. The forces which the complete structure, or certain areas of it, experiences at high speed due to air resistance are dependent on how close the aircraft's speed approaches that of sound. Since the speed of sound depends on atmospheric pressure and density, it will vary with altitude, and this suggests that in order to fly an aircraft within safe speed limits a different air speed would have to be maintained for each altitude. This obviously is not acceptable and it therefore became necessary to have a means whereby the ratio of the aircraft's speed, V, and the speed of sound, a, could be computed from pressure measurement and indicated in a conventional manner. This ratio, V/a, is termed the *Mach number* (M), a parameter which has now assumed great significance in practical aerodynamics, and the instrument which measures the ratio $M = V/a$ is termed a *Machmeter*. Before dealing with the construction and operation of the instrument we will review briefly some of the effects encountered by an aircraft flying at high speed.

The passage of an aircraft through the air sets up vibratory disturbances of the air which radiate from the aircraft in the form of pressure or sound waves. At speeds below that of sound, termed *subsonic* speeds, these waves radiate away from the aircraft in much the same way as ripples move outward from a point at which a stone is thrown into water. When speeds approach that of sound, however, there is a drastic change in the sound-wave radiation pattern. The aircraft is now travelling almost as fast as its own sound waves and they begin to pile up on one another ahead of the aircraft, thus increasing the air resistance and setting up vibrations of the air causing turbulence and buffeting of the aircraft, thus imposing severe stresses. Shock waves are also developed which cause a breakdown in the airflow over wings, fuselage and tail unit and thus lead to stalling and difficulties in movement of control surfaces. For regular flying in this *transonic* range, and at *supersonic* speeds, aircraft must be designed accordingly; for example, looking at some of to-day's high-speed aircraft, we note that the wings and horizontal stabilizer are swept back; this delays the onset of shock waves and thus permits a greater airspeed or Mach number to be attained.

It is possible for buffeting to occur when an aircraft is flying at speeds below that of sound because the location, aerodynamic shape and profile of parts of the structure may allow the airflow over them to reach or exceed *sonic* speed. The Mach number at which this occurs is referred to as the *critical* Mach

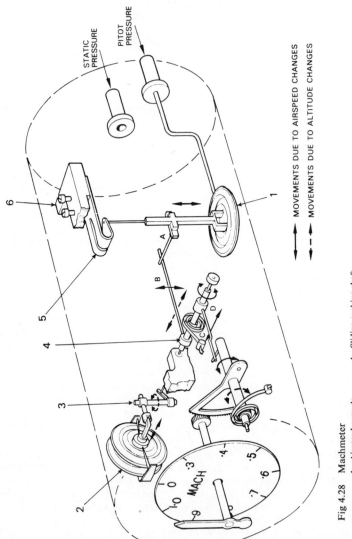

Fig 4.28 Machmeter

1 Airspeed capsule
2 Altitude capsule
3 Altitude rocking shaft
4 Sliding rocking shaft
5 Calibrating spring } square-law compensation
6 Calibration screws

number (M_{crit}) of that particular aircraft, and being a ratio of airspeed and sonic speed, it will be the same for any altitude.

A Machmeter is a compound flight instrument which accepts two variables and uses them to compute the required ratio. The construction of the instrument is shown in Fig 4.28.

The first variable is *airspeed* and therefore a mechanism based on the conventional airspeed indicator is adopted to measure this in terms of the pressure difference $p - p_s$, where p is the total or pitot pressure and p_s the static pressure. The second variable is *altitude*, and this is also measured in the conventional manner, i.e. by means of an aneroid capsule sensitive to the static pressure, p_s. Deflections of the capsules of both mechanisms are transmitted to the indicator pointer by rocking shafts and levers, the dividing function of the altitude unit being accomplished by an intermediate sliding rocking shaft.

Let us assume that the aircraft is flying under standard sea-level conditions at a speed V of 500 m.p.h. The speed of sound at sea-level is approximately 760 m.p.h., therefore the Mach number is $500/760 = 0.65$. Now, the speed measured by the airspeed mechanism is, as we have already seen, equal to the pressure difference $p - p_s$, and so the sliding rocking shaft and levers A, B, C and D will be set to angular positions determined by this difference. The speed of sound cannot be measured by the instrument, but since it is governed by static pressure conditions, the altimeter mechanism can do the next best thing and that is to measure p_s and feed this into the indicating system, and thereby setting a datum position for the point of contact between the levers C and D. Thus a Machmeter indicates the Mach number ratio V/s in terms of the pressure ratio $(p - p_s)/p_s$, and for the speed and altitude conditions assumed the pointer will indicate 0.65.

What happens at altitudes above sea-level? As already pointed out, the speed of sound decreases with altitude, and if an aircraft is flown at the same speed at all altitudes, it gets closer to and can exceed the speed of sound. For example, the speed of sound at 10,000 ft decreases to approximately 650 m.p.h., and if an aircraft is flown at 500 m.p.h. at this altitude, the Mach number will be $500/650 = 0.75$, a 10% increase over its sea-level value. It is for this reason that critical Mach numbers (M_{crit}) are established for the various types of high-speed aircraft, and being constant with respect to altitude it is convenient to express any speed limitations in terms of such numbers.

We may now consider how the altitude mechanism of the Machmeter functions in order to achieve this, by taking the case of an aircraft having an M_{crit} of, say, 0.65. At sea-level and as based on our earlier assumption, the measured airspeed would be 500 m.p.h. to maintain $M_{crit} = 0.65$. Now, if the aircraft is to climb to and level off at a flight altitude of 10,000 ft, during the climb the decrease of static pressure p_s causes a change in the pressure ratio. It affects the pressure difference $p - p_s$ in the same manner as a conventional indicator is affected, i.e. the measured airspeed is decreased. The airspeed mechanism therefore tends to make the pointer indicate a lower Mach number. However, the altitude mechanism simultaneously responds to the decrease in p_s, its capsule expanding and causing the sliding rocking shaft to carry lever C towards the pivot point of lever D.

The magnification ratio between the two levers is therefore altered as the altitude mechanism divides $p - p_s$ by p_s, lever D being forced down so as to make the pointer maintain a constant Mach number of 0.65.

The critical Mach number for a particular type of aircraft is indicated by a pre-adjusted lubber mark located over the dial of the Machmeter

MAXIMUM SAFE AIRSPEED INDICATION

In some types of high-performance aircraft, indicators are used which combine the functions of a Machmeter and a conventional airspeed indicator, and present on a common dial the indicated airspeed and the maximum safe airspeed compatible with altitude. It consists of two measuring elements as in a Machmeter, but as can be seen from Fig 4.29, they are independent of each other and drive separate indicating elements.

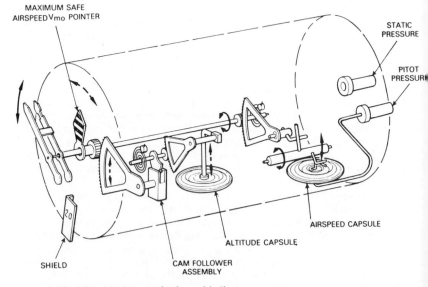

Fig 4.29   Maximum safe airspeed indicator

The airspeed indicating element is of the split-pointer type similar to that described in Chapter 2. Both pointers are held in contact with each other by a hairspring (not shown in the diagram) and they rotate in unison over the outer scale up to 200 knots. At this speed, movement of the rear pointer is arrested by a combined shield and stop so that indications above 200 knots are taken over by the front pointer registering against an inner scale. The maximum safe airspeed over a predetermined range of altitude and airspeed is indicated by a single yellow and black striped pointer called the *velocity maximum operating* (V_{mo}) *pointer*. It registers against the inner scale only and is rotated by the altitude capsule from a starting position appropriate to the instrument type and operating ranges of the aircraft in which it is to be installed. The position is set during calibration of the instrument by means of the cam follower assembly, which is dimensioned to give the altitude and airspeed relationship over the required maximum speed range.

A lubber mark is provided at the periphery of the dial and can be positioned at any of the outer scale markings by means of a setting knob at the bottom left-hand corner of the bezel. It enables a pilot to preset a speed indication which requires monitoring during a particular operating phase (e.g. the aircraft lift-off speed during take-off).

AIRSPEED SWITCH UNITS

Airspeed switch units, like their altitude counterparts, can be used for a variety of warning applications; for example, in aircraft fitted with a fatigue meter, a switch unit is employed to switch the meter on and off at predetermined airspeeds; a unit may also be used to operate an audio signal device and so give warning of an overspeed. Whatever the application, the switch units are special adaptations of conventional airspeed indicator mechanisms.

A typical unit is shown in Fig 4.30 (a). The switch mechanism is housed in a standard airspeed indicator case and consists of a capsule, a set of low-speed and high-speed contacts and a relay. A direct-current power supply of 28 V is fed to the circuit via a plug at the rear of the case.

When the differential pressure across the capsule increases it expands until at a predetermined pressure it closes the low-speed contacts. At a slightly higher predetermined differential pressure, the high-speed contacts close and complete the circuit to the relay coil thus energizing it. Operation of the relay then completes the external circuit to the fatigue meter or other device connected to the switch unit. In some applications, particularly those involving audible warning devices, only one predetermined speed is required; in such cases, the circuit is arranged so that only the low-speed contact or the high-speed contact completes the external circuit.

Figure 4.30 (b) shows another type of airspeed switch which basically employs the same number of components as the one described in the preceding paragraphs. The essential differences are the casing construction, single contacts instead of double, plunger contact assembly instead of leaf contacts, and an external scale for adjustment of the contact setting.

Combined Airspeed Indicator and Warning Switch

Figure 4.31 is a schematic of an airspeed indicator incorporating an undercarriage position warning system. The system comes into operation when the approach speed of the aircraft is reached and the undercarriage is not extended and locked in position. If this should happen, a warning flag visible through an aperture in the dial, adjacent to the approach speed graduations, commences to oscillate.

The system comprises a pair of contacts actuated by the capsule, and connected to an operating coil and a relay. The relay, which is supplied with 28 V d.c. via the aircraft's undercarriage "down" lock system, controls the operating coil which, in turn, actuates the warning flag by means of the rocking shaft assembly.

When the undercarriage is retracted, the direct-current supply is applied to the indicator, but while airspeeds are above the preset value, the capsule holds the contacts open. As the airspeed decreases, the capsule contracts until at the preset value the contacts close and complete two parallel circuits. One circuit energizes the relay and the other charges a capacitor. Energizing of the relay

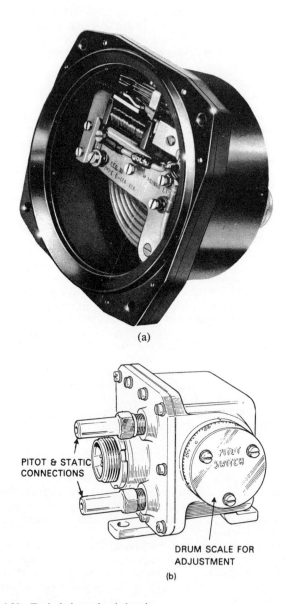

Fig 4.30 Typical airspeed switch units
(a) Low-speed and high-speed contact type   (b) Single-contact type

Pitot-static Instruments and Systems 97

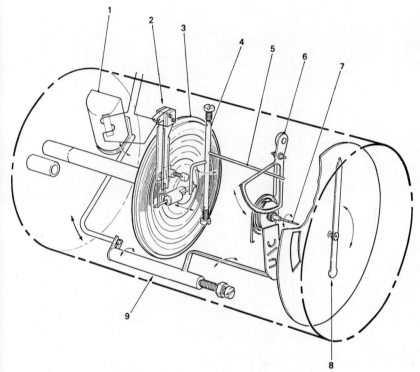

Fig 4.31 Combined airspeed indicator and warning system

1 Flag operating coil
2 Contact assembly
3 Capsule
4 Rocking shaft
5 Bimetal arm
6 Sector
7 Handstaff
8 Pointer
9 Rocking shaft flag assembly

causes its contacts to change over, thus interrupting the supply to the relay coil and also connecting a supply to the flag operating coil, causing the flag to move into the dial aperture. At the same time the capacitor is supplied with direct current and it starts charging. When it discharges it does so through the relay coil and holds the contacts in position until the discharge voltage reaches a point at which it is insufficient to hold the relay energized. The contacts change over once again and de-energize the flag actuator coil, causing the flag to move away from the dial aperture. The cycle is then repeated and at such a frequency that the flag appears in the aperture at approximately half-second intervals.

VERTICAL SPEED INDICATORS

These indicators also known as rate-of-climb indicators, are the third of the primary group of pitot-static flight instruments, and are very sensitive differential pressure gauges, designed to indicate the rate of altitude change from the change of static pressure alone.

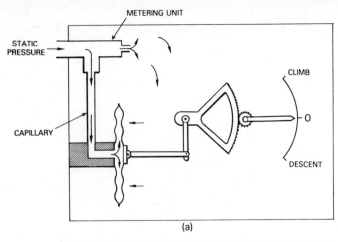

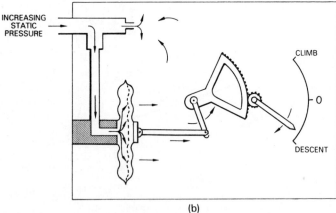

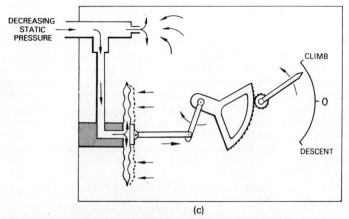

Fig 4.32 Principle of vertical speed indicator

    (a) Level flight: zero differential pressure across capsule
    (b) Aircraft descending: metering unit maintains case pressure lower than capsule pressure, changing it at the same rate and thereby creating a constant differential pressure across the capsule
    (c) Aircraft climbing: metering unit creates a constant differential pressure across capsule by maintaining case pressure higher than capsule pressure

Now, it may be asked why employ a differential pressure gauge which requires two pressures to operate it when there is only one pressure really involved? Why not use an altimeter since it too measures static pressure changes? These are fair enough questions, but the operative clause is "the rate at which the static pressure changes," and as this involves a time factor we have to introduce this into the measuring system as a pressure function. It is accomplished by using a special air metering unit, and it is this which establishes the second pressure required.

An indicator consists basically of three main components, a capsule, an indicating element and a metering unit, all of which are housed in a sealed case provided with a static pressure connection at the rear. The dial presentation is such that zero is at the 9 o'clock position; thus the pointer is horizontal during straight and level flight and can move from this position to indicate climb and descent in the correct sense. Certain types of indicator employ a linear scale, but in the majority of applications indicators having a mechanism and scale calibrated to indicate the logarithm of the rate of pressure change are preferred. The reason for this is that a logarithmic scale is more open near the zero mark and so provides for better readability and for more accurate observation of variations from level flight conditions.

The indicator mechanism is shown in schematic form in Fig 4.32, from which it will be noted that the metering unit forms part of the static pressure connection and is connected to the interior of the capsule by a length of capillary tube. This tube serves the same purpose as the one employed in an airspeed indicator, i.e. it prevents pressure surges reaching the capsule. It is, however, of a greater length due to the fact that the capsule of a vertical speed indicator is much more flexible and sensitive to pressure. The other end of the metering unit is open to the interior of the case to apply static pressure to the exterior of the capsule.

Let us now see how the instrument operates under the three flight conditions: (a) level flight, (b) descent, and (c) climb.

In level flight, air at the prevailing static pressure is admitted to the interior of the capsule, and also to the instrument case through the metering unit. Thus, there is zero differential across the capsule and the pointer indicates zero.

We will now consider the operation during a descent. At the instant of commencing the descent the differential pressure will still be zero, but as the aircraft descends into the higher static pressure this will be applied at the static pressure connection of the instrument causing air to flow into the capsule and case.

As the capsule is directly connected to the static pressure connection, the flow of air will create the same pressure inside the capsule as that prevailing at the levels through which the aircraft is descending. The pressure inside the case, however, is not going to be the same because the metering unit is a specially calibrated leak assembly designed to restrict the flow of air into or out of the instrument case. Therefore, as far as the case pressure is concerned, it is still at the same value which obtained at the original level flight altitude, and cannot build up at the same rate as the pressure in the capsule is increasing. The restriction of the metering unit thus provides the second pressure from one source and establishes a differential pressure across the capsule, causing it to distend and make the pointer indicate a descent. This, of course is just what is required, but during the descent the case pressure must be

100 *Aircraft Instruments*

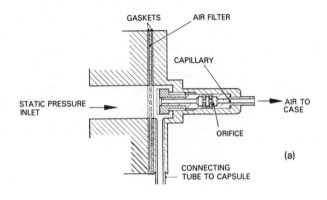

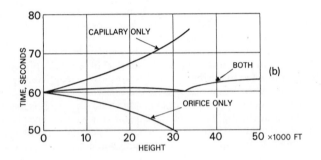

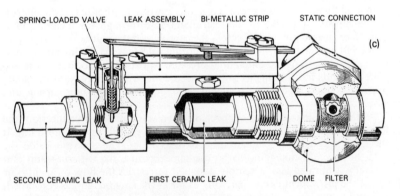

Fig 4.33 Vertical speed indicator metering units
  (*a*) Capillary and orifice type
  (*b*) Capillary and orifice characteristics
  (*c*) Ceramic type

maintained lower than the capsule pressure and made to change at the same rate in order to obtain a constant differential pressure.

The metering unit being a restrictor, increases the velocity of the air flowing into the static pressure connection, and as happens with devices of this nature, increased velocity brings about a reduction in pressure. In addition, the instrument case is of much greater volume than the capsule; consequently the flow of air into the case is going to take some time to build up a pressure equal to that coming in at the static pressure connection. By the time this is reached, however, the aircraft will have descended to a new altitude and the static pressure will have again changed. Thus, the metering unit introduces the required rate and time-lag factors, and differential pressure across the capsule which positions the pointer to indicate the altitude change in feet per minute. The design of a system is such that it takes approximately four seconds for the case pressure to build up to that in the capsule; but as the capsule always has an unrestricted air flow to it, it will lead the case by four seconds and so there will be a constant difference in pressure between them corresponding to four seconds in time. The differential pressures produced are not very large, a typical value being approximately 20 mm H_2O at full-scale deflection of the pointer.

During a climb, the metering unit will establish the required factors and differential pressure, but as the static pressure under this condition is a decreasing one, and because the metering unit restricts the flow out of the case, the case pressure leads the capsule pressure.

Apart from the changes of static pressure with changes of altitude, which as we know are not constant, air temperature, density and viscosity changes are other very important variables which must be taken into account, particularly as the instrument depends on rates of air flow. From the theoretical and design standpoints, a vertical speed indicator is therefore quite complicated, but the metering units are designed to compensate for the effects of variables over the ranges normally encountered.

The construction and operation of two typical units are described in the following paragraphs.

Metering Units

The unit shown in Fig 4.33 (*a*) is known as a "capillary-and-orifice" type, the two devices in combination providing compensation for the effects of the atmospheric pressure and temperature variables, as shown at (*b*).

The pressure difference across a capillary, for a constant rate of climb, increases with increasing altitude and at a constant temperature. Thus, the use of a capillary alone would introduce a positive error in instrument indications at altitudes above sea-level. With an orifice, the effect is exactly the opposite. The primary reasons for the difference are that the air flow through a capillary is a laminar one while that through an orifice is turbulent; furthermore, the rate of flow through a capillary varies directly as the differential pressure, while that through an orifice varies as the square root of the differential pressure. In combining the two devices we can therefore obtain satisfactory pressure compensation at a given temperature.

The differential pressure across a capillary also depends on the viscosity of the air, and as this is proportional to the absolute temperature, it therefore decreases with decreasing temperature. The differential pressure across an orifice varies inversely as the temperature, and therefore increases with decreasing temperature. Thus, satisfactory temperature compensation can be obtained by combining the two devices. The sizes of the orifice and capillary are chosen so that the readings of the indicator will be correct over as wide a range of temperature and altitude conditions as possible.

The second unit, illustrated in Fig 4.33 (c), is known variously as the "ceramic type," and "porous-pot type," and is a little more complicated in its construction, because a mechanical temperature/viscosity compensator is incorporated.

It will be noted that the air from the static connection flows into the capsule via a capillary tube, into the case via two ceramic porous tubes, and also through the valve of the viscosity compensator. The valve opening is controlled by the effects of temperature on a bimetallic strip.

Under low altitude conditions the effect of temperature on the bimetallic strip is such that the valve is open by a certain amount, so that after flowing through the first ceramic tube the air passes into the case via the open valve. At higher altitudes the static air fed to the instrument is at a lower temperature, and because its viscosity decreases with decreasing temperature, the overall effect is to reduce the pressure differential and give rise to errors. However, the lower temperature also has its effect on the bimetallic strip, causing it to bend and close the valve. The air must now also flow through the second ceramic tube in order to get into the case, and as they are in series and of calibrated porosity, the differential pressure is increased and maintained. In practice, the valve takes up positions between open and closed, but the calibration of the metering unit as a whole gives a constant differential pressure for a fixed rate of climb or descent at any altitude.

Typical Indicator

The construction of a typical vertical speed indicator employing an orifice and capillary type of metering unit is shown in Fig 4.34. It consists of a cast aluminium-alloy body which forms the support for all the principal components with the exception of the metering unit, which is secured to the rear of the indicator case. Displacements of the capsule in response to differential pressure changes are transmitted to the pointer via a link and rocking-shaft magnifying system, and a quadrant and pinion. The magnifying system and indicating element are balanced by means of an adjustable weight attached to the rocking shaft. The flange of the metering unit connects with the static pressure connection of the indicator case, and it also acts as a junction for the capillary tube.

Range setting of the instrument during initial and subsequent calibrations is achieved by two calibration springs which bear on a stem connected to the centre-piece of the capsule. The purpose of these springs is to exert forces on the capsule and so achieve the correct relationship between the capsule's pressure/deflection characteristics and the pointer position at all points of the scale. The forces are controlled by two rows of screws, located in a calibration bracket, which vary the effective length of their respective springs. The upper row of screws and the upper spring control the rate of descent calibration, while the lower row of screws and lower spring control the rate of climb.

A feature which meets a common requirement for all types of vertical speed indicator is adjustment of the pointer to the zero graduation. The form taken by the adjustment device depends on the instrument design, but in the mechanism we have been considering, it consists of an eccentric shaft coupled by a gearwheel to a pinion on a second shaft which extends to the bottom centre of the bezel. The exposed end of the shaft is provided with a screwdriver slot. When the shaft is rotated the eccentric shaft is driven round to displace a plate bearing against the eccentric. The plate is also in contact with the underside of the capsule, and as a result the capsule is moved up or down, the movement being transferred to the pointer via the magnifying system and pointer gearing. The range of pointer adjustment around zero depends on the climb and descent range of the instrument but ± 200 and ± 400 ft/min are typical values.

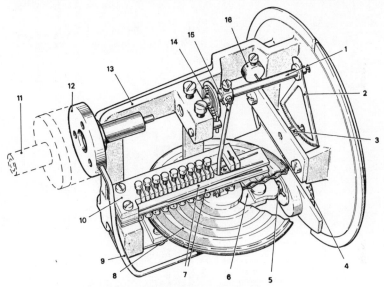

Fig 4.34 Typical vertical speed indicator mechanism

1 Rocking shaft assembly
2 Sector
3 Handstaff pinion
4 Gearwheel
5 Eccentric shaft assembly
6 Capsule plate assembly
7 Calibration springs
8 Capsule
9 Capillary tube
10 Calibration bracket
11 Static connection
12 Metering unit
13 Mechanism body
14 Hairspring
15 Link
16 Balance weight

AIR DATA SYSTEMS

In certain types of aircraft currently in service, indications of airspeed, Mach number and altitude are provided by instruments which, unlike their conventional counterparts, depend on electric signals for their operation. In conjunction with a central computer unit they form part of an *air data system* inter-connected in the manner shown schematically in Fig 4.35.

The computer, which basically is of the electro-mechanical analogue type, receives the pressure data from the aircraft's pitot-static system, computes this in the form of electric signals and transmits the signals via a synchronous link and servo system to the indicator. Other air data requirements, e.g

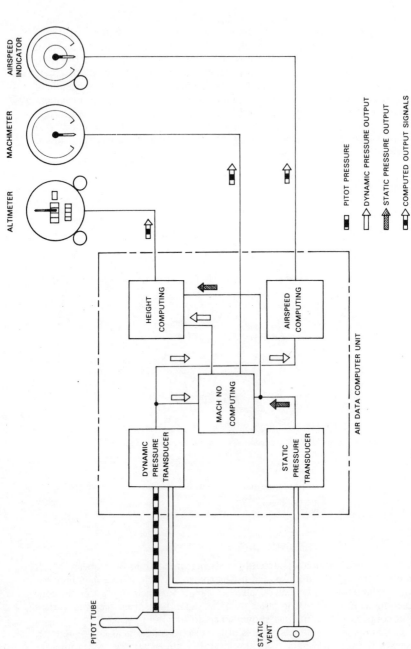

Fig 4.35 Schematic of an air data system

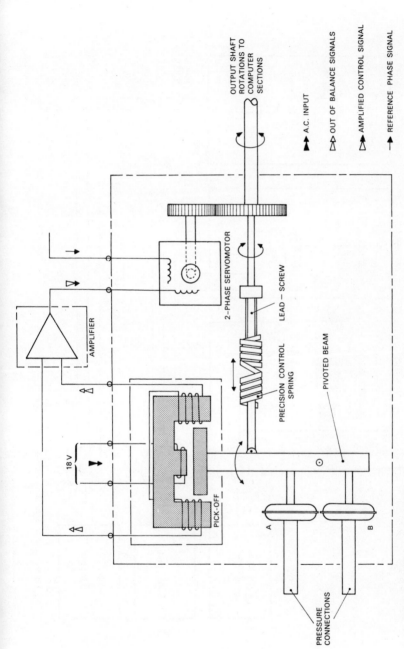

Fig 4.36 Pressure transducer

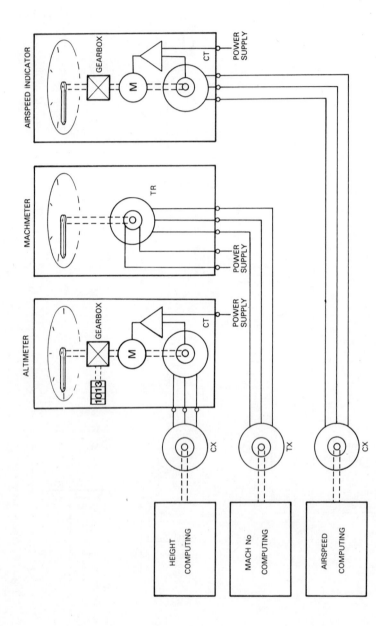

Fig 4.37 Air data signal transmission system

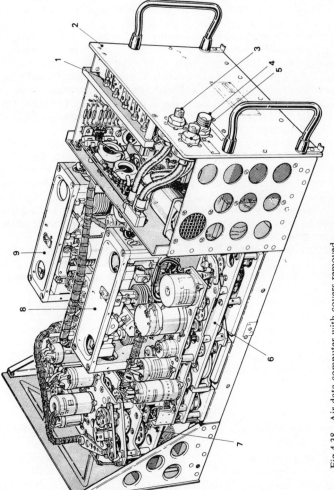

Fig 4.38 Air data computer with covers removed
1 Amplifier bank
2 Front panel
3 Pitot connector
4 Static (speed) connector
5 Static (height) connector
6 Rate of climb unit
7 Mach unit
8 Static pressure transducer
9 Pitot-static transducer

signals for an automatic flight control system, are catered for by additional output signal circuits.

The sensing of the pressures is accomplished by two pressure transducers the components of which are shown in Fig 4.36. The two units are identical except that in the dynamic pressure transducer, capsule A is supplied with pitot pressure while in the static pressure transducer it is evacuated and sealed.

When a change of pressure takes place the capsules respond and produce a force which causes a slight deflection of the pivoted beam. The I-bar of the inductive pick-off unit is also displaced relative to the limbs of the E-shaped bar, causing an out-of-balance signal to be induced in the coils. The signal is amplified and applied to the control phase of the servo motor which drives the output shaft and a lead-screw. The lead-screw is coupled to the pivoted beam via a precision control spring so that, as the screw rotates, the spring tension is varied. When the spring tension balances the force due to the pressure change, the beam is returned to its central position to reduce the inductive pick-off signal to zero, thus stopping the servo motor.

The output shaft rotates at a speed proportional to the pressure applied; thus in the dynamic pressure transducer the speed is proportional to the difference between pitot and static pressures, while in the static pressure transducer it is proportional only to static pressure. The shafts of both transducers are mechanically coupled with cam and servo mechanisms the function of which is to allow for pressure error by "shaping" mechanical and electrical signals. The mechanisms are contained within computing sections appropriate to the measurements of airspeed, Mach number and altitude.

The corrected output signals from the three computing sections are then transmitted mechanically to the rotors of synchros linked to receiving units which control pointer movements of the corresponding flight instruments (Fig. 4.37). A control transmitter CX and transformer CT form the synchronous link for airspeed and altitude indications, while simple torque synchros TX, TR are employed for the indication of Mach number.

Correction for pressure error (see page 62) is obtained by using cams in the computing mechanisms, calibrated and profiled to suit the pitot and static pressure conditions of a particular aircraft.

A complete air data computer is shown in Fig 4.38.

QUESTIONS

4.1 What are the principal components and instruments which comprise an aircraft pitot-static system?

4.2 Draw a line diagram of a dual pitot and static system for port and starboard instrument panels in an aircraft. (SLAET)

4.3 (a) Explain the principle of pitot pressure measurement and how the $\frac{1}{2}\rho V^2$ law is derived.

(b) Define the law to which current types of airspeed indicator are calibrated.

4.4 Sketch and describe the construction of a pressure head. (SLAET)

4.5 How are the effects of turbulent air passing through the static slots of a pitot-static tube neutralized?

4.6 Draw the circuit diagram of a typical pressure head heating system and explain its operation.

4.7 What effects do the drain holes of a pressure head have on the indications of the instruments connected to it?

4.8 (a) What is meant by the pressure error of a pitot-static tube?
(b) How are its effects minimized?
(c) Explain why a vertical speed indicator is unaffected by pressure error.

4.9 Define the following: (i) troposphere, (ii) tropopause, (iii) stratosphere.

4.10 State two units which are commonly used in atmospheric pressure measurements. (SLAET)

4.11 What will be the effect on the density of an air mass if the temperature decreases but the pressure remains constant? (SLAET)

4.12 (a) What do you understand by the term "standard atmosphere"?
(b) State the assumptions made by the I.C.A.O. Standard.

4.13 With the aid of a diagram explain the operating principle of an aneroid barometer.

4.14 Describe the construction and operation of an altimeter. Explain any special features which improve its accuracy. (SLAET)

4.15 Discuss problems of the misreading of altimeters in association with the instrument presentation. How do designers overcome these problems: (a) in aircraft fitted with older types of instrument, (b) by re-design and replacement of altimeters. (SLAET)

4.16 What is the difference between "pressure altitude" and "indicated altitude"?

4.17 Explain how an altimeter is compensated for errors due to atmospheric pressure changes.

4.18 Define the three principal Q codes used for altimeter pressure settings.

4.19 (a) For what purpose are altimeter switches provided in aircraft?
(b) With the aid of a diagram, describe the construction and operation of a typical unit.

4.20 With the aid of a sketch, describe the construction and explain the operation of a servo altimeter. State its advantages over a sensitive altimeter. (SLAET)

4.21 (a) Define the $\frac{1}{2}\rho V^2$ law as it is applied to present-day airspeed indicators.
(b) State the values inserted in the formula for an indicator calibrated in knots.

4.22 Describe the construction and operation of an airspeed indicator. Explain how it is compensated for environmental conditions. (SLAET)

4.23 (a) What is meant by "square-law compensation" of an airspeed indicator?
(b) With the aid of a diagram explain the operation of a typical device.

4.24 Define the following: (i) Mach number, (ii) critical Mach number.

4.25 Describe how Mach number is indicated by measuring in terms of the ratio $p - p_s/p_s$.

4.26 Describe how the functions of a Machmeter and an airspeed indicator are combined to give an indication of maximum safe airspeed.

4.27 With the aid of a sketch, describe the construction of a vertical speed indicator. (SLAET)

4.28 Explain the operation of a vertical speed indicator when the aircraft in which it is installed goes from a level flight attitude into a climb attitude.

4.29 With the aid of a diagram explain the construction and operation of a metering unit incorporating a viscosity compensator valve.

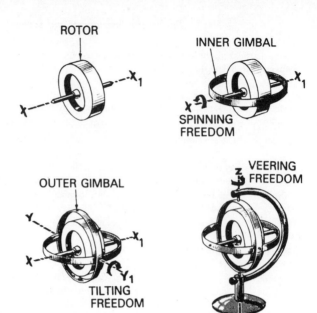

Fig 5.1 Elements of a gyroscope

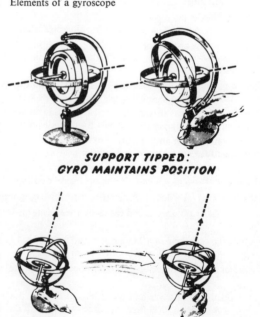

Fig 5.2 Gyroscopic rigidity

5 Primary Flight Instruments (Attitude Indication)

This chapter deals with the two flight instruments which provide a pilot with the necessary indications of the pitch, bank and turn attitudes of his aircraft. As both these instruments and the instruments covered in Chapters 6 and 7 are dependent on gyroscopic properties, this subject forms the opening to the present chapter.

THE GYROSCOPE AND ITS PROPERTIES

As a mechanical device a *gyroscope* may be defined as a system containing a heavy metal wheel, or rotor, universally mounted so that it has three degrees of freedom: (i) *spinning freedom* about an axis perpendicular through its centre (axis of spin XX_1); (ii) *tilting freedom* about a horizontal axis at right angles to the spin axis (axis of tilt YY_1); and (iii) *veering freedom* about a vertical axis perpendicular to both the spin and tilt axes (axis of veer ZZ_1).

The three degrees of freedom are obtained by mounting the rotor in two concentrically pivoted rings, called inner and outer gimbal rings. The whole assembly is known as the *gimbal system* of a *free* or *space gyroscope*. The gimbal system is mounted in a frame as shown in Fig 5.1, so that in its normal operating position, all the axes are mutually at right angles to one another and intersect at the centre of gravity of the rotor.

The system will not exhibit gyroscopic properties unless the rotor is spinning; for example, if a weight is hung on the inner gimbal ring, it will merely displace the ring about axis YY_1 because there is no resistance to the weight. When the rotor is made to spin at high speed the device then becomes a true gyroscope possessing two important fundamental properties: *gyroscopic inertia* or *rigidity*, and *precession*. Both these properties depend on the principle of conservation of angular momentum, which means that the angular momentum of a body about a given point remains constant unless some force is applied to change it. *Angular momentum* is the product of the moment of inertia (I) and angular velocity (ω) of a body referred to a given point—the centre of gravity in the case of a gyroscope.

If a weight is now hung on the inner gimbal ring with the rotor running, it will be found that the gimbal ring will support the weight, thus demonstrating the first fundamental property of rigidity. However, it will also be found that the complete gimbal system will start rotating about the axis ZZ_1, such rotation demonstrating the second property of precession. Figure 5.2 illustrates how gyroscopic rigidity may be demonstrated. If the frame and outer gimbal ring are tipped about the axis YY_1, the gyroscope maintains its spin axis in the horizontal position. If the frame is either rotated about the axis 22_1 or is swung in an arc, the spin axis will continue to point in the same direction.

These rather intriguing properties can be exhibited by any system in which a rotating mass is involved. Although it was left for man to develop gyroscopes

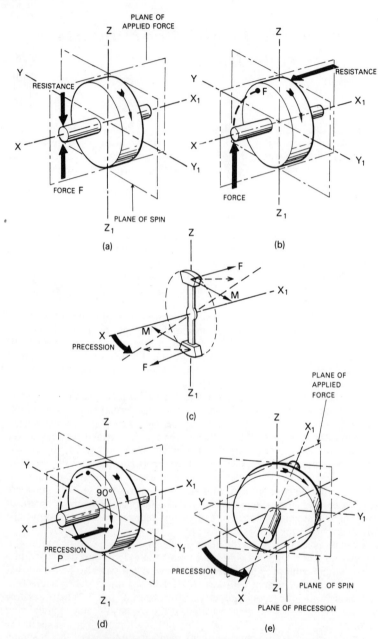

Fig 5.3 Gyroscopic precession
(a) Gyro resists force
(b) Transmission of force
(c) Effect on rotor segments
(d) Generation of precession
(e) Effect of precession

Primary Flight Instruments (Attitude Indication)

and associated devices, it is true to say that gyroscopic properties are as old as the earth itself: it too rotates at high speed and so possesses rigidity, and although it has no gimbal system or frame on which external forces can act, it can, and does, precess. There are, however, many mechanical examples around us every day and one of them, the bicycle, affords a very simple means of demonstration. If we lift the front wheel off the ground, spin it at high speed, and then turn the handlebars, we feel rigidity resisting us and we feel precession trying to twist the handlebars out of our grasp. The flywheel of a motor-car engine is another example. Its spin axis is in the direction of motion of the car, but when turning a corner its rigidity resists the turning forces set up, and as this resistance always results in precession, there is a tendency for the front of the car to move up or down depending on the direction of the turn. Other familiar examples are aircraft propellers, compressor and turbine assemblies of jet engines: gyroscopic properties are exhibited by all of them.

The two properties of an actual gyroscope may be more closely defined as follows:

Rigidity. The property which resists any force tending to change the plane of rotation of its rotor. This property is dependent on three factors: (i) the mass of the rotor, (ii) the speed of rotation, and (iii) the distance at which the mass acts from the centre, i.e. the radius of gyration.

Precession. The angular change in direction of the plane of rotation under the influence of an applied force. The change in direction takes place, not in line with the applied force, but always at a point 90° away in the direction of rotation. The rate of precession also depends on three factors: (i) the strength and direction of the applied force, (ii) the moment of inertia of the rotor, and (iii) the angular velocity of the rotor. The greater the force, the greater is the rate of precession, while the greater the moment of inertia and the greater the angular velocity, the smaller is the rate of precession.

Precession of a rotor will continue, while the force is applied, until the plane of rotation is in line with the plane of the applied force and until the directions of rotation and applied force are coincident. At this point, since the applied force will no longer tend to disturb the plane of rotation, there will be no further resistance to the force and precession will cease.

Determining the Direction of Precession

The direction in which a gyroscope will precess under the influence of an applied force may be determined by means of vectors and by solving certain gyrodynamic problems, but for illustration and practical demonstration purposes, there is an easy way of determining the direction in which a gyroscope will precess and also of finding out where a force must be applied for a required direction of precession. It is done by representing all forces as acting directly on the rotor itself.

At (*a*) in Fig 5.3, the rotor of a gyroscope is shown spinning in a clockwise direction and with a force, *F*, applied upwards on the inner gimbal ring. In transmitting this force to the rim of the rotor, as will be noted from (*b*), it will act in a horizontal direction. Let us imagine for a moment that the rotor is broken into segments and concern ourselves with two of them at opposite sides of the rim as shown at (*c*). Each segment has motion *m* in the direction of

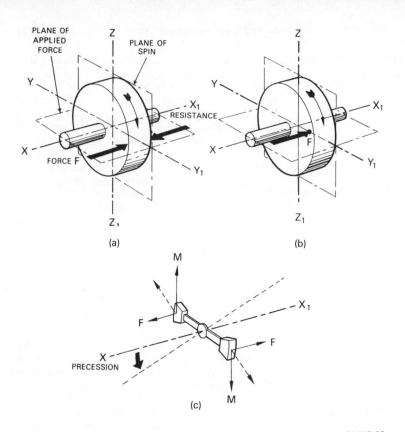

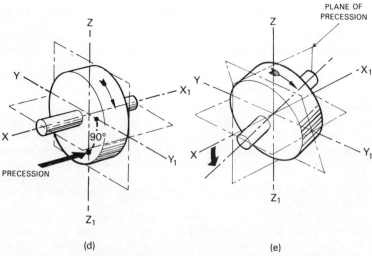

Fig. 5.4. Gyroscopic precession
(a) Gyro resists force
(b) Transmission of force
(c) Effect on rotor segments
(d) Generation of precession
(e) Effect of precession

rotor spin, so that when the force F is applied there is a tendency for each segment to move in the direction of the force. As the gyroscope possesses rigidity this motion is resisted, but the segments will turn about the axis ZZ_1 so that their direction of motion is along the resultant of motion m and force F. The other segments will be affected in the same way; therefore, when they are all joined to form the solid mass of the rotor it will precess at an angular velocity proportional to the applied force (see diagrams (d) and (e)).

In the example illustrated in Fig 5.4 (a), a force, F, is shown applied on the outer ring; this is the same as transmitting the force on the rotor rim at the point shown in diagram (b). As in the previous example this results in the direction of motion changing to the resultant of motion m and force F_1. This time, however, the rotor precesses about the axis YY_1 as indicated at (d) and (e).

REFERENCES ESTABLISHED BY GYROSCOPES

For use in aircraft, gyroscopes must establish two essential reference datums: a vertical flight reference against which pitch and roll attitude changes may be detected, and a directional reference against which changes about the vertical axis may be detected. These references are established by gyroscopes having their spin axes arranged vertically and horizontally respectively, as shown in Fig 5.5.

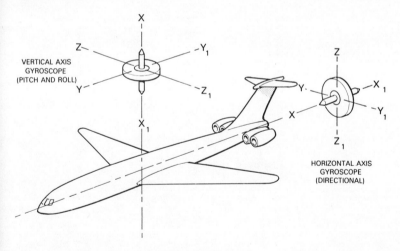

Fig 5.5 References established by a gyroscope

Both types of gyroscope utilize the fundamental properties in the following manner: rigidity establishes a stabilized reference unaffected by movement of the supporting body, and precession controls the effects of the earth's rotation, bearing friction, unbalance, etc., thus maintaining the reference in the required position.

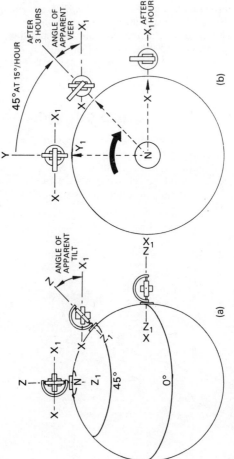

Fig 5.6 Effects of earth's rotation and curvature
(a) Apparent tilt (b) Apparent veer

Limitations of a Free Gyroscope

Aircraft in flight are still very much a part of the earth, i.e. all references must be with respect to the earth's surface. The free or space gyroscope we have thus far considered would, however, serve no useful purpose and must be corrected for the effects of the earth's rotation and curvation.

EFFECTS OF EARTH'S ROTATION AND CURVATURE

As the earth rotates, the spin axis of a gyroscope maintains its position in space in the same way as when the frame of the gyroscope is tipped or turned by hand. Thus, if it is imagined that a free gyroscope is positioned at the equator and is started with its spin axis horizontal in an EW direction as shown in Fig 5.6 (a), then to an observer out in space, the spin axis would appear to maintain its starting position, but to an observer on the earth, the spin axis would appear to tilt at a rate corresponding to the earth's angular velocity of 15°/hour. *Apparent tilt* would also be observed if the gyroscope was fixed in an aeroplane and carried, say, from the North Pole to the South Pole regardless of the line of longitude the aeroplane followed.

Figure 5.6 (b) shows a plan view of the earth, with a space gyroscope set up at the North Pole. An observer standing at a point opposite one end of the rotor axis will, after a period of a few hours, note that he is looking at some other part of the gyroscope. This *apparent veer*, or drift, is due to the fact that the observer is carried round the gyroscope by the earth's rotation. At the Equator, and because the gyroscope would have appeared to tilt until its axis was pointing toward the centre of the earth, there would be no apparent veer. Veering can only take place about the axis ZZ_1, and at the Equator this axis is aligned with that of the spin axis XX_1, and because they would both rotate with the earth, no apparent movement would be observed about the axis ZZ_1.

The Earth or "Tied" Gyroscope

It will be appreciated from the foregoing that at any intermediate position on the earth, the gyroscope would experience both apparent veer and apparent tilt, dependent upon the latitude. Therefore, for a complete rotation of the earth the two ends of the gyroscope axis would appear to move in circles and the gyroscope as a whole would appear to make a conical movement. If the gyroscope is carried in an aircraft, the same conical movement will be traced, but the angular velocity of this movement will be decreased or increased depending on whether the EW component of the aircraft's speed is towards east or west. The NS component of the speed will increase the maximum divergence of the gyroscope axis from the vertical, the amount of divergence depending on whether the aircraft's speed has a North or South component and also on whether the gyroscope is situated in the Northern or Southern hemisphere.

Before a space gyroscope can be of practical use it must be controlled so that it precesses to maintain its plane of spin relative to the earth; in other words, it must be "tied" to the earth. When the gyroscope is "tied" by a gravity control it is then called an *earth gyroscope*, and in Fig 5.7 the effect of this control is compared with that of a space gyroscope under the same conditions. At position 1 both gyroscopes are positioned at the North Pole with their axes pointing towards the earth's centre. At position 2 they have both been moved to a point midway between the North Pole and the Equator. It will be noted that the axis of the earth gyroscope has moved from its original position in space, its axis continuing to point to the earth's centre, whilst the space gyroscope maintains its position in space. At the Equator the earth gyroscope has been controlled so that its axis still points to the earth's centre, but the axis of the space gyroscope is still horizontal to the earth's surface.

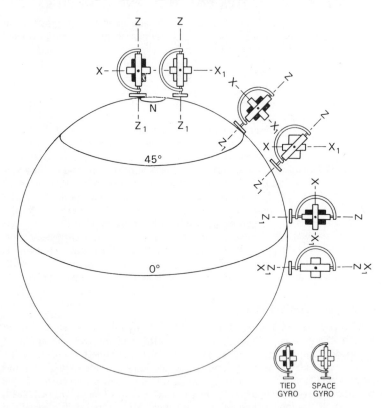

5.7 Tied and space gyroscopes at various latitudes

METHODS OF OPERATING GYROSCOPIC FLIGHT INSTRUMENTS

There are two principal methods used for driving the rotors of gyroscopic flight instruments: pneumatic and electric. In the pneumatic method the case of an instrument is connected to either an engine-driven vacuum pump, or a venturi located externally and in the slipstream of a propeller. The pump, or venturi, creates a vacuum which is regulated by a relief valve at between 3·5 and 4·5 in Hg. Certain types of turn-and-slip indicator operate at a lower value, and this is obtained by an additional regulating valve in the indicator supply line.

Each instrument has two connections: one is made to the pump or venturi line, and the other is made internally to a spinning jet system and is open to the surrounding atmosphere. When vacuum is applied to the instruments, the pressure within their cases is reduced to allow the surrounding air to enter and emerge through the spinning jets. The jets are adjacent to "buckets" cut in the periphery of each instrument rotor so that the jet stream turns the rotors at high speed.

At high altitudes vacuum-driven gyroscopic instruments suffer from the effects of a decrease in vacuum due to the lower atmospheric pressure; the resulting reduction in rotor speeds affecting gyroscopic stability. Other disadvantages of vacuum operation are weight due to pipelines, special arrangements to control the vacuum in pressurized cabin aircraft, and, since air must pass through bearings, the possibility of contamination by corrosion and dirt particles.

To overcome these disadvantages and to meet instrumentation demands for high-performance aircraft, gyroscopic instruments were designed for operation from aircraft electrical systems. In current applications this applies particularly to gyro horizons and turn-and-slip indicators; electrically driven directional gyros form part of remote-indicating compass systems, the principles of which are dealt with in Chapter 7.

The power supplies generally used are 115V, 400Hz, 3-phase current derived from an inverter or engine-driven alternator, and 28V direct current, the latter being required for the operation of some types of turn-and-slip indicator. The gyroscopes of alternating-current instruments utilize the principle of the squirrel-cage induction motor, and because the frequency of the power supply is high, greater rotor speeds (of the order of 24,000 rev/min.) are possible, thus providing greater rigidity and stability of indications. The design of direct-current operated gyroscopes is based on the principle of the conventional permanent-magnet type of motor.

THE GYRO HORIZON

The *gyro horizon*, or *artificial horizon* as it is sometimes called, indicates the pitch and bank attitude of an aircraft relative to the vertical, and for this purpose employs a gyroscope whose spin axis is maintained vertical by a gravity-sensing device, so that effectively it serves the same purpose as a pendulum but with the advantage that aircraft attitude changes do not cause it to oscillate.

Indications of pitch and bank attitude are presented by the relative positions of two elements, one symbolizing the aircraft itself, and the other in the form of a bar stabilized by the gyroscope and symbolizing the natural horizon. Supplementary indications of bank are presented by the position of a pointer, also gyro-stabilized, and a fixed bank angle scale. Two methods of presentation are shown in Fig 5.8.

As noted elsewhere, gyro horizons may be pneumatically or electrically operated, but in both cases the application of the gyroscopic principles is the same and may be understood by referring to Fig 5.9. The gimbal system is arranged so that the inner ring forms the rotor casing, and is pivoted parallel to the aircraft's athwartships axis YY_1; and the outer ring is pivoted parallel

120 Aircraft Instruments

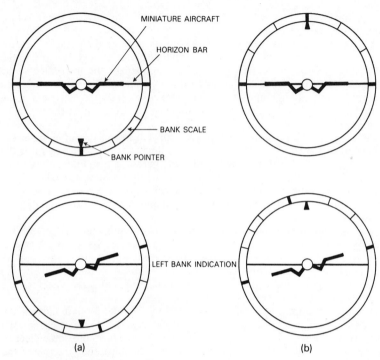

Fig 5.8 Gyro horizon presentations
(a) Bottom bank scale   (b) Top bank scale

to the aircraft's fore-and-aft axis ZZ_1. The outer ring pivots are located at the front and rear ends of the instrument case. The element symbolizing the aircraft may be either rigidly fixed to the case, or externally adjusted up and down for pitch trim setting.

In operation the gimbal system is stabilized so that in level flight the three axes are mutually at right angles. When there is a change in the aircraft's attitude, it goes into a climb say, the instrument case and outer ring will turn about the axis YY_1 of the stabilized inner ring.

The horizon bar is pivoted at the side and to the rear of the outer ring, and engages an actuating pin fixed to the inner ring, thus forming a magnifying lever system. In a climb attitude the bar pivot carries the rear end of the bar upwards causing it to pivot about the stabilized actuating pin. The front end of the bar and the pointer therefore move downwards through a greater angle than that of the outer ring, and since movement is relative to the symbolic aircraft element, a climbing attitude is indicated.

Changes in the lateral attitude of the aircraft, i.e., banking, turn the instrument case about the axis ZZ_1 and the whole stabilized gimbal system. Hence, lateral attitude changes are indicated by movement of the symbolic aircraft element relative to the horizon bar, and also by relative movement between the bank angle scale and the pointer.

Primary Flight Instruments (*Attitude Indication*) 121

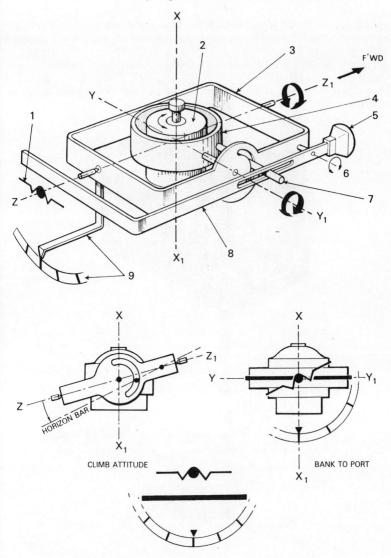

Fig 5.9 Principle of gyro horizon

Freedom of gimbal system movement about the roll and pitch axes is 360° and 85° respectively, the latter being restricted by means of a "resilient stop." The reason for restricting the pitch movement of a gyro horizon to 85° is to prevent "gimbal lock." This is a phenomenon which can occur when the rotor spin axis coincides with the outer gimbal ring axis, i.e. inner ring turned through 90°. Under such conditions the gyroscope no longer has two degrees of freedom, and if it is turned, a torque will be applied to the inner ring to precess the outer ring into a continuous spin. Once spinning has begun, the inner ring remains locked to the outer ring regardless of the attitude assumed by the gyroscope thereafter.

The use of stops presents another problem. When they contact the outer ring it precesses through 180° about its axis, such motion being known as "tumbling." However, it is very seldom that the instrument pitch limits are exceeded under normal operating conditions.

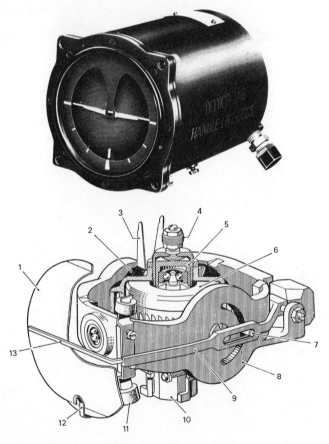

Fig 5.10 Vacuum-driven gyro horizon

Vacuum-driven Gyro Horizon

A typical version of a vacuum-driven gyro horizon is shown in Fig 5.10. The rotor is pivoted in ball bearings within a case forming the inner ring, which in turn is pivoted in a rectangular-shaped outer ring. The lower rotor bearing is fitted into a recess in the bottom of the rotor casing, whereas the upper bearing is carried in a housing which is spring-loaded within the top cap to compensate for the effects of differential expansion between the rotor shaft and case under varying temperature conditions.

A background plate which symbolizes the sky is fixed to the front end of the outer ring and carries the bank pointer which registers against the bank-angle scale.

The outer ring has complete freedom through 360° about the roll axis. A "resilient stop" limiting the ±85° pitch movement is fitted on the top of the rotor casing.

The horizon bar and pointer are an accurately balanced assembly pivoted in plain bearings on the side of the outer ring and slotted to engage the actuating pin projecting from the rotor case. Pitch attitude changes are indicated by the pointer set at right angles to the bar and positioned in front of the "sky plate." For those instruments having a complete roll freedom gimbal system, the horizon bar pointer is in two parts so as to prevent it fouling the front pivot of the outer gimbal ring.

In the rear end cover of the instrument case, a connection is provided for the coupling of the vacuum supply. A filtered air inlet is also provided in the cover and is positioned over the outer-ring rear-bearing support and pivot, which are drilled to communicate with a channel in the outer ring. This channel terminates in diametrically-opposed spinning jets within the rotor casing, the underside of which has a number of outlet holes drilled in it.

With the vacuum system in operation, a depression is created so that the surrounding atmosphere enters the filtered inlet and passes through the channels to the jets. The air issuing from the jets impinges on the rotor buckets, thus imparting even driving forces to spin the rotor at approximately 15,000 rev/min in an anticlockwise direction as viewed from above. After spinning the rotor, the air passes through a pendulous vane unit attached to the underside of the rotor casing, and is finally drawn off by the vacuum source.

The purpose and operation of the pendulous vane unit is described under "Erection Systems for Gyro Horizons" on page 125.

Electric Gyro Horizon

A typical electric gyro horizon is shown in Fig 5.11; as will be noted, it is made up of the same basic elements as the vacuum-driven type, with the exception that the vertical gyroscope is a 3-phase squirrel-cage induction motor (consisting of a rotor and a stator).

One of the essential requirements of any gyroscope is to have the mass of the rotor concentrated as near to the periphery as possible, thus ensuring maximum inertia. This presents no difficulty where solid metal rotors are concerned, but when adopting electric motors as gyroscopes some rearrangement of their basic design is necessary in order to achieve the desired effect. An induction motor normally has its rotor revolving inside the stator, but to make one small enough to be accommodated within the space available would mean too small a rotor mass and inertia. However, by designing the rotor and its bearings so that it rotates on the outside of the stator, then for the same required size of motor the mass of the rotor is concentrated further from the centre, so that the radius of gyration and inertia are increased. This is the method adopted not only in gyro horizons but in all instruments and systems employing electric gyroscopes.

The motor assembly is carried in a housing which forms the inner gimbal

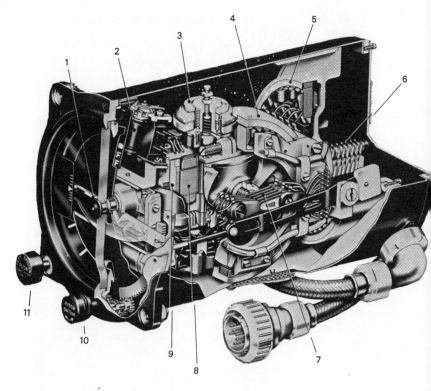

Fig 5.11 Electric gyro horizon

1. Miniature aircraft
2. Power-failure indicator assembly
3. Gyro assembly
4. Roll-torque motor
5. Pitch-torque motor
6. Slip-ring assembly
7. Gyro-gimbal contact assembly
8. Stator
9. Rotor
10. Pitch-trim adjusting knob
11. Fast-erection push switch

ring supported in bearings in the outer gimbal ring, which is in turn supported on a bearing pivot in the front cover glass and in the rear casting. The horizon bar assembly is in two halves pivoted at the rear of the outer gimbal ring and is actuated in a manner similar to that already described on page 120.

The 115 V 400 Hz 3-phase supply is fed to the gyro stator via slip rings, brushes and finger contact assemblies. The instrument employs a torque-motor erection system, the operation of which is described on page 127.

When power is switched on a rotating magnetic field is set up in the gyro stator which cuts the bars forming the squirrel-cage in the rotor, and induces a current in them. The effect of this current is to produce magnetic fields around the bars which interact with the stator's rotating field causing the rotor to turn at a speed of approximately 20,000–23,000 rev/min. Failure of the power supply is indicated by a flag marked OFF and actuated by a solenoid.

ERECTION SYSTEMS FOR GYRO HORIZONS

As in the case of directional gyroscopes, the vertical gyroscope of a gyro horizon can also drift as a result of bearing friction, earth's rotation, and movement of the aircraft over the earth's surface. They must therefore be provided with a controlling device to erect and maintain their rotor axes in the vertical position.

The systems adopted depend on the particular design of gyro horizon, but they are all of the gravity-sensing type and in general fall into two main categories: mechanical and electrical. The construction and operation of some systems which are typical of those in current use are described on this and the following pages.

Mechanical Systems

Pendulous Vane Unit

This unit, employed with the air-driven instrument described on page 123, is shown in detail in Fig 5.12. It is fastened to the underside of the rotor housing and consists of four knife-edged pendulously suspended vanes clamped in pairs on two intersecting shafts supported in the unit body. One shaft is parallel to the axis YY_1 and the other parallel to the axis ZZ_1 of the gyroscope. In the sides of the body there are four small elongated ports, one under each vane.

The air, after having spun the gyro rotor, is exhausted through the ports, emerging as four streams: one forward, one rearward and two athwartships. The reaction of the air as it flows through the ports applies a force to the unit body. The vanes, under the influence of gravity, always hang in the vertical position, and it is this feature which is utilized to govern the airflow from the ports and to control the forces applied to the gyroscope by the air reaction.

When the gyroscope is in its normal vertical position as shown in Fig 5.12 (b), the knife-edges of the vanes bisect each of the ports (A, B, C and D), making all four port openings equal. This means that all four air reactions are equal and the resultant forces about each axis are in balance.

If now the gyroscope is displaced from its normal vertical position, for example, its top is tilted towards the front of the instrument as at (C); the pair of vanes on the axis YY_1 remain vertical, thus opening the port (D) on the right-hand side of the body and closing that (B) on the left. The increased reaction of the air from the open port results in a torque being applied to the body in the direction of the arrow, about axis XX_1.

This torque is equivalent to one applied on the underside of the rotor and to the left, or at the top of the rotor at point F as shown at (d). As a gyroscope rotor always moves at a point 90° away from the point of an applied torque, then in this case the rotor is precessed at point P back to the vertical when the vanes again bisect the ports to equalize the air reactions.

Ball-type Erection Unit

This unit utilizes the precessional forces resulting from the effects of gravity on a number of steel balls displaced within a rotating holder suspended from the gyro housing.

126 Aircraft Instruments

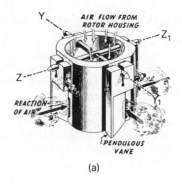

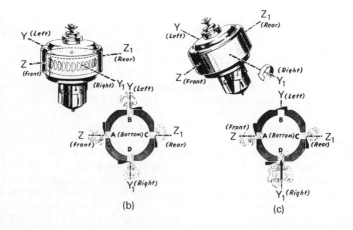

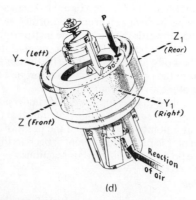

Fig 5.12 Pendulous vane erection unit
(a) Construction
(b) Precession due to air reaction
(c) Gyro in vertical position
(d) Gyro tilted

The holder of the *ball erector mechanism* encloses from five to eight balls, the number depending upon the particular design, which are free to roll across a radiused erecting disc. A plate having a number of specially profiled hooks is fixed around the inner edge of the holder. The spacing of the hooks is chosen so as to regulate the release of the balls when the gyroscope tilts, and to shift their mass to the proper point on the erecting disc to apply the force required for precession. Rotation of the holder takes place through reduction gearing from the gyro rotor shaft, the speed of the holder being 25 rev/min.

When the gyroscope is in its normal operating position as shown in Fig 5.13 (*a*), the balls change position as the holder rotates but their mass remains concentrated at the centre of the erecting disc. Under this condition, gravity exerts its greatest pull at the centre of the mass, and therefore all forces about the principal axes of the gyroscope are in balance.

At (*b*) the gyro vertical axis is shown displaced about pitch axis YY_1 away from the front of the instrument. The displacement of the ball erector mechanism causes the balls to roll towards the hooks, which at that instant are on the low side; therefore the force due to gravity is now shifted to this side. Since the hooked plate is rotating (clockwise viewed from above), the balls and the point at which the force is acting will be carried round to the left-hand side of the ball holder. In this position the balls remain hooked and their mass remains concentrated to allow a torque to be exerted at the left-hand side of the ball holder as indicated at (*c*). This torque may also be considered as acting directly on the left-hand bearing of the gyro housing and outer ring. Transferring this point of applied torque to the rim of the rotor precession will then take place at a point $90°$ ahead in the direction of rotation. As may be seen from the diagram, the gyro housing will now start precessing about the axis YY_1 to counteract the displacement.

As the erector mechanism continues to rotate, the balls will be carried round to the high side of the holder, but one by one they will roll into the hooks at the lower side. Thus, their mass is once more concentrated at this side allowing the torque and precession to be maintained as they are carried around to the left-hand side. This action continues with diminishing movement of the balls as the gyroscope erects to its normal operating position, at which the balls are at the centre of the disc and the force due to gravity is again concentrated at the centre of the mass.

Displacement of a gyroscope in other directions about its lateral or longitudinal axis will result in actions similar to those described, and it is left to the reader as a useful exercise to determine the forces, torques and precession produced by both types of erection system and for a chosen displacement.

Torque Motor and Levelling Switch System

This system is used in a number of electrically-operated gyro horizons and consists of two torque control motors independently operated by mercury levelling switches, which are mounted, one parallel to the athwartships axis, and the other parallel to the fore-and-aft axis. The disposition of the torque motors and switches is illustrated diagrammatically in Fig 5.14.

The athwartships mounted switch detects displacement of the gyroscope in roll and is connected to its torque motor so that a corrective torque is applied around the pitch axis. Displacement of the gyroscope in pitch is detected by

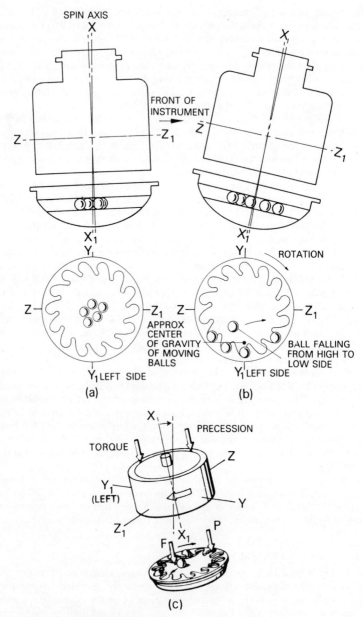

Fig 5.13 Ball-type erection unit
  (a) Gyro vertical
  (b) Gyro tilted towards front of instrument
  (c) Precession to the vertical

the fore-and-aft mounted levelling switch, which is connected to its torque motor so that corrective torques are around the roll axis.

Each levelling switch is in the form of a sealed glass tube containing three electrodes and a small quantity of mercury. They are mounted in adjustable cradles set at right angles to each other on a switch block positioned beneath the gyro housing. The tubes are filled with an inert gas to prevent arcing at the electrodes as the mercury makes contact and also to increase the rupturing capacity.

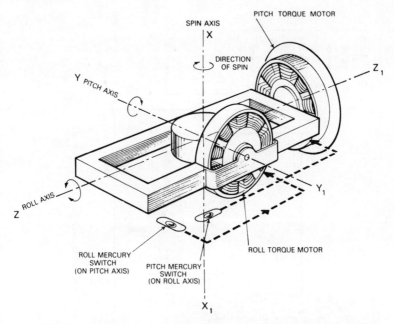

Fig 5.14 Arrangement of a torque-motor and levelling-switch erection system
Pitch torque motor: rotor fixed to rear part of casing; stator fixed to outer ring
Roll torque motor: rotor fixed to outer ring; stator fixed to gyro housing

The torque motors comprise a squirrel-cage-type laminated-iron rotor mounted concentrically about a stator, the iron core of which has two windings, one providing a constant field and called the "reference winding," and the other in two parts so as to provide a reversible field, and called the "control winding." Both windings are powered from a step-down auto-transformer connected between phases A and B of the 115 V supply to the gyro horizon.

The electrical interconnection of all the components comprising the system is indicated in Fig 5.15.

When the gyro is running and in its normal operating position, the mercury in the levelling switches lies at the centre of the tubes and is in contact with the centre electrode. The two outer electrodes, which are connected across the control windings of the torque motor stators, remain open. The auto-transformer reduces the voltage to a selected value (typically 20 V) which is then fed

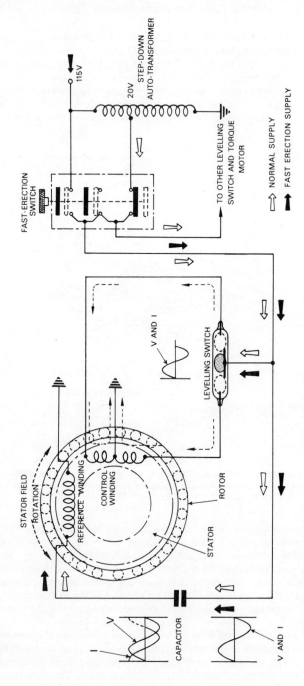

Fig 5.15 Circuit diagram of torque-motor and levelling-switch erection system

to the centre electrode of the switches and the reference windings of the torque motors. Thus, in the normal operating position of the gyroscope, current flows through the reference windings only.

Let us consider what happens when the gyroscope is displaced about one of its axes, to the front of the instrument, say, and about the pitch axis YY_1. The pitch-levelling switch will also be displaced and the mercury will roll to the forward end of the tube to make contact with the outer electrode. This completes a circuit to one part of the control winding of the pitch torque motor causing current to flow through it in the direction indicated in Fig 5.15. The stator of the roll torque motor will also be displaced inside its fixed rotor, but will receive no current at its control winding since the roll-levelling switch is unaffected by displacement about the pitch axis.

The necessary corrective torque to the gimbal system must be applied by the pitch torque motor, and in order to do this, the magnetic field of its stator must be made to rotate. The voltage applied to the reference winding is fed via a capacitor, and in any alternating-current circuit containing capacitance, the phase of the current is shifted so as to lead the voltage by 90°. In the circuit to the control winding there is no capacitance; therefore, the voltage and current in this winding are in phase, and since the reference and control windings are both fed from the same source, then the reference winding current must also lead that in the control winding by 90°. This out-of-phase arrangement, or *phase quadrature*, applies also to the magnetic field set up by each winding.

Thus, with current and flux flowing through the control winding in the direction resulting from the gyro displacement considered, a resultant magnetic field is produced which rotates in the stator in an anticlockwise direction.

As the field rotates, it cuts the closed-circuit bar-type conductors of the squirrrel-cage rotor causing a current to be induced in them. The effect of the induced current is to produce magnetic fields around the bars which interact with the rotating field in the stator creating a tendency for the rotor to follow the stator field.

This tendency is immediately opposed because the rotor is fixed to the instrument case; consequently, a reactive torque is set up in the torque motor which is exerted at the rear bearing of the outer ring. We may consider this torque as being exerted at a point on the gyro rotor itself so that precession will take place at a point 90° ahead in the direction of rotation. This precession will continue until the gyro and mercury switch are once again in the normal operating position.

It will be clear from Fig 5.15 that displacement of the gyroscope in the opposite direction will cause current to flow in the other part of the levelling-switch control winding, thus reversing the direction of the stator magnetic field and the resulting precession.

Fast-erection Systems

In some types of electrically-operated gyro horizon employing the torque-motor method of erection, the arrangement of the levelling switches is such that, if the gyro rotor axis is more than 10° from the vertical, the circuits to the torque motors are interrupted so that the gimbal system will never erect. For example, in one design a commutator switch, known as a bank erection cut-out, is carried on the outer gimbal ring about the roll axis, and serves to reduce erection errors during turns involving bank angles greater than 10°, by

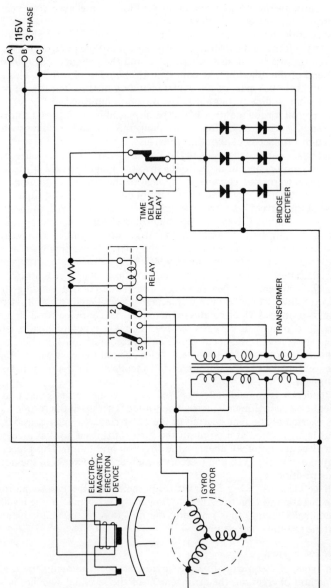

Fig 5.16 Circuit diagram of electromagnetic method of fast erection

opening the circuits to both levelling switches. Thus, if on resuming level flight the gimbal system has not remained accurately stabilized so as to be within the 10° angle, the erection cut-out will maintain the erection system in the inoperative condition.

Furthermore, it is possible for the gyroscope of a gyro horizon to have "toppled," or to be out of the vertical by too great an angle prior to starting the instrument; then due to the low erection rate of the system normally adopted, it would take too long before the required accuracy of indication was obtained.

In order, therefore, to overcome these effects and to bring the gyroscope to its normal operating position as quickly as possible, a fast erection system may be provided. Two typical systems in current use are described in the following paragraphs.

Fast-erection Switch
This method is quite simple in operation. The switch (Fig 5.15) consists of several contacts connected in the power supply lines to the erection-system torque motors and levelling switches.

Under normal operating conditions of the gyro horizon, the switch remains spring-loaded to the "off" position and the low-voltage supply from the auto-transformer passes over one closed contact of the switch to the erection system, the other contacts remaining open.

Whenever the gyroscope goes beyond the appropriate angular limits, the erection system circuit must be restored and the gyroscope's position brought back to normal as quickly as possible. This is achieved by pushing in the switch so that the contact in the low-voltage supply line opens to isolate the erection system from the auto-transformer, and the upper contacts close. The closure of these contacts completes the circuit to the torque motors and levelling switches, but the power supply to them is now changed over from the low-voltage value to the full line voltage of 115 V from one of the phases. This results in an increase of current through the stator windings of the torque motors, and the greater torque so applied increases the erection rate from the normal value of 5° per minute to between 120° and 180° per minute, depending on the particular design.

There are two important precautions which have to be observed when using this switch. Firstly, the switch must not be depressed for longer than 15 seconds to prevent overheating of the stator coils due to the higher current. The second precaution is one to be observed under flight conditions: the switch must only be depressed during straight and level steady flight and/or shallow angles of climb or descent. If acceleration or deceleration forces are present, the gyroscope will precess and produce false indications of pitch and bank attitude.

Electromagnetic Method of Fast Erection
In this method, a circular-shaped electromagnet is secured to the inside of the instrument case above an umbrella-shaped armature mounted on the gyro rotor housing. The armature is of approximately the same diameter as the magnet.

Control of the electromagnet and the erection time is achieved by an auxiliary power control unit containing a three-phase transformer, bridge rectifier, thermally-operated time-delay relay and a standard d.c. relay, all interconnected as shown in the circuit diagram of Fig 5.16.

When the normal 115 V alternating-current supply is initially switched on, it is fed to contacts 1 and 2 of the standard relay, and from one phase, through the time-delay relay, to the bridge rectifier. The direct curren tobtained from the rectifier is then supplied to the coil of the electromagnet, which, on being energized, produces a magnetic field radiating symmetrically from a small centre pole to a circular outer pole.

If, at the moment of switching on the power supply the gyro rotor housing and hence the armature are tilted away from the centre of the magnet, then the magnetic field is no longer symmetrical with respect to the centre of the armature. Under these conditions, therefore, a greater force is exerted on one side of the armature than the other, and the applied torque is in such a direction as to cause the gyro housing to erect to the vertical and bring the top of the armature into line with the centre of the magnet before the rotor is up to full speed.

In addition to passing through the electromagnet, the direct current from the rectifier also passes through the coil of the standard relay which is thus energized at the same time as the electromagnet. The resulting changeover of the relay contacts causes the 115 V supply to be fed to the tapping points 2 and 3 on the transformer primary winding. This has the effect of reducing the number of turns of the winding; in other words, the transformer is of the step-up type, the voltage of the secondary winding in this particular application being increased to 185 V.

After approximately 20 seconds, the time-delay relay opens and disconnects the direct current from the electromagnet. The standard relay then de-energizes and switches the gyro rotor circuit from the transformer to the normal 115 V supply, the rotor running up to full speed some seconds later.

Erection Rate

This is the term used to define the time taken, in degrees per minute, for the rotor axis of a vertical gyroscope to take up its vertical position under the action of its gravity-sensing erection system.

For the ideal gyro horizon, the erection rate should be as fast as possible under all conditions, but in practice such factors as speed, turning and acceleration of the aircraft, and earth's rotation all have their effect and must be taken into account. During turns the erection system is acted upon by centrifugal forces and is displaced to make the gyro follow it by precession. Therefore, the maximum erection rate that can be used is limited by the maximum error that can be tolerated during turns. The minimum rate is governed by the earth's rotation, speed of the aircraft, and random changes of precession due to bearing friction, variations in rotor speed, and gimbal system unbalance.

Thus, erection systems must be designed so that, for small angular displacements of the rotor axis from the vertical, the erecting couple is proportional to the displacement, while for larger displacements it is made constant. It is also arranged that the couple gives equal erection rates for any rotor axis displacement in any direction in order to reduce the possibility of a slow cumulative error during manoeuvres.

Normal erection rates provided by some typical erection systems are 8° per minute for vacuum-driven gyro horizons and from 3° to 5° per minute for electrically-driven gyro horizons.

Primary Flight Instruments (*Attitude Indication*)   135

Errors Due to Acceleration and Turning

As we have already learned, the erection devices employed in gyro horizons are all of the pendulous gravity-controlled type. This being so, it is possible for them to be displaced by the forces acting during the acceleration and turning of an aircraft, and unless provision is made to counteract them the resulting torques will precess the gyro axis to a false vertical position and so present a false indication of an aircraft attitude. For example, let us consider the effects of a rapid acceleration in the flight direction, firstly on the vane type of erection device and secondly on the levelling-switch and torque-motor type (see Fig 5.17).

The force set up by the acceleration will deflect the two athwartships-mounted vanes to the rear, thus opening the right-hand port. The greater reaction of air flowing through the port applies a force to the rotor and the torque causes it to precess forward about the axis YY_1. The horizon bar is thus displaced upwards, presenting a false indication of a descent.

With the levelling-switch and torque-motor type of erection device, the acceleration force will deflect the mercury in the pitch levelling switch to the rear of the glass tube. A circuit is thus completed to the pitch torque motor which also precesses the gyroscope forward and displaces the horizon bar to indicate a descent.

In both cases the precession is due to a natural response of the gyroscope, and the pendulous vanes and the mercury always return to their neutral positions, but for so long as the disturbing forces remain, such positions apply only to a false vertical. When the forces are removed the false indiction of descent will remain initially and then gradually diminish under the influence of precession restoring the gyro axis to its normally true vertical.

It should be apparent from the foregoing that, during periods of deceleration, a gyro horizon will present a false indication of a climb.

When an aircraft turns, false indications about both the pitch and bank axes can occur due to what are termed "gimballing effects" brought about by forces acting on both sets of pendulous vanes and both levelling switches. There are, in fact, two errors due to turning; erection errors and pendulosity errors.

Erection Errors

As an aircraft enters a correctly banked turn, the gyro axis will initially remain in the vertical position and an accurate indication of bank will be presented. In this position, however, the fore-and-aft mounted pendulous vanes, or roll levelling switch, are acted upon by centrifugal force. The gyroscope will therefore be subjected to a torque applied in such a direction that it tends to precess the gyro axis towards the aircraft perpendicular along which the resultant of centrifugal and gravity forces is acting. Thus, the gyroscope is erected to a false vertical and introduces an error in bank indication.

An analysis of the error can be made with the aid of Fig 5.18, which illustrates the case of an aircraft turning to starboard through 360° from a starting point A. The centrifugal force experienced by the gyro axis in the false vertical position during the turn is constant and at right angles to the instantaneous heading. This means that when the aircraft changes its heading at a constant rate during a 360° turn, the top of the gyro axis will trace out a circular path

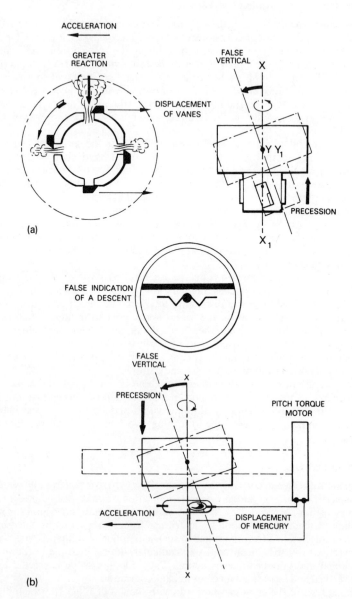

Fig 5.17 Acceleration error
(a) Vane-type erection system
(b) Levelling-switch and torque-motor erection system

which is 90° in advance of the aircraft heading. The circle at the left of Fig 5.18 represents the path of the gyroscope axis, and any chord of this circle will indicate the tilt of the axis in relation to the true vertical. The chord AB', for example, represents the direction of tilt after the aircraft has turned through 90°. In relating this tilt to the gyroscope and the response of its gravity-controlled erection devices to the turn, it can be resolved into two components, one forward and the other to starboard. Thus, in addition to an error in bank indication an error in pitch is presented when the aircraft is at point B of its turn. In a similar manner, the chord AC' indicates the direction of tilt after 180°; at this point the tilt is maximum and the bank error has been reduced to zero, leaving maximum error in pitch indication. The direction of tilt after

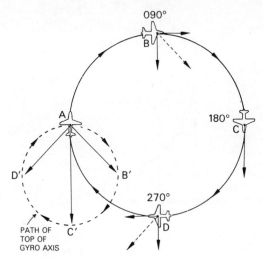

Fig 5.18   Erection error

270° is indicated by chord AD', and resolving this into its two components as at point D, we see that the pitch error is the same as at B but the bank error is in the opposite direction. On returning to point A the tilt of the gyro axis would be zero.

Compensation for Erection Errors

Erection errors may be compensated by one of the following three methods: (i) inclination of the gyro spin axis, (ii) erection cut-out, and (iii) pitch-bank compensation.

Inclined Spin Axis

The method of inclining the spin axis is based on the idea that, if the top of the axis can describe a circle about itself during a turn, then only a single constant error will result. In its application, the method is mechanical in form and varies with the type of gyro horizon, but in all cases the result is to impart a constant forward (rearward in some instruments) tilt to the gyro axis from the true vertical. The angle of tilt varies but is usually either 1·6° or 2·5°. In vacuum-driven types the athwartships-mounted pendulous vanes are balanced so that the gyroscope is precessed to the tilted position; in certain electric gyro horizons the pitch mercury switch is fixed in a tilted position so that the gyroscope is precessed away from the true vertical in order to overcome what it detects as a pitch error. The linkages between gyroscope and horizon bar are so arranged that during level flight the horizon bar will indicate this condition.

The effect of the tilt is shown in Fig 5.19, where point A represents the end

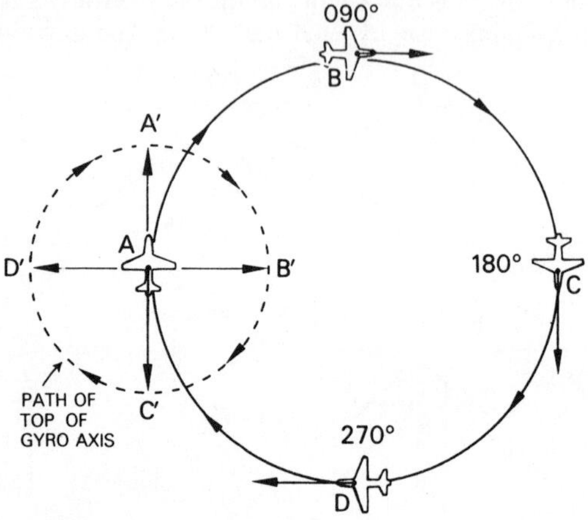

Fig 5.19 Compensated erection error

of the true vertical through the centre of the rotor, and AA' represents the direction of tilt (forward in this case). During a turn to starboard the top of the gyro axis describes a circle about point A at the same rate as the aircraft changes heading. The amount of tilt and its direction in relation to the aircraft during the turn are therefore constant.

Erection Cut-out

The erection cut-out method is one applied to certain types of electric gyro horizon and operates automatically whenever the aircraft banks more than 10° in either direction. It consists basically of a commutator made up of a conducting segment and an insulated segment, and two contacts or brushes connected in series with the bank levelling switch. The commutator is located on the bank axis at the rear of the outer gimbal ring, and in the straight and level flight condition the two brushes bear against the conducting segment thus completing the power supply circuit to the levelling switch.

During a turn there is relative movement between the commutator and brushes due to banking, and when the angle of bank exceeds 10° the insulated segment comes under one or other of the brushes and so interrupts the supply to the bank levelling switch. Displacement of the mercury by centrifugal force cannot therefore energize the relevant torque motor and cause precession to a false vertical.

Owing to the function of the cut-out, no erection about the bank axis is possible if the power supply is switched on to the instrument when its gimbal system is tilted more than 10° about this axis. However, the supply can be connected by means of a "fast erection" circuit which by-passes the cut-out in the manner described on page 133.

"Pitch-bank Erection"

The third method, generally referred to as "pitch-bank erection," is a combined one in which the bank levelling switch is disconnected during a turn and its erection system is controlled by the pitch levelling switch. It is intended to correct the varying pitch and bank errors and operates only when the rate of

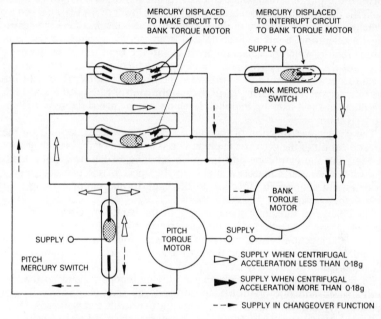

Fig 5.20 "Pitch-bank" erection

turn causes a centrifugal acceleration exceeding $0.18g$, which is equivalent to a 10° tilt of the bank erection switch. The system is shown schematically in Fig 5.20, and from this we note that two additional mercury switches, connected as a double-pole changeover switch, are provided and are interconnected with the normal pitch and bank erection systems.

Let us consider first a turn to the left and one creating a centrifugal acceleration less than $0.18g$. In such a turn, the mercury in the bank levelling switch will be displaced to the right and will bridge the gap between the supply electrode and the right-hand electrode, thus completing a circuit to the bank torque motor. This is the same as if the gyro axis had been tilted to the right at the commencement of the turn; the bank torque motor will therefore precess the gyro to a false vertical, left of the true one. At the same time, the gyro axis tilts forward due to gimballing effect, and the mercury in the pitch levelling switch, being unaffected by centrifugal acceleration, moves forward and completes a circuit to the pitch torque motor, which precesses the gyro rearwards. The two curved changeover switches, which are also mounted about the bank axis, do not come into operation since the mercury in each switch is not displaced sufficiently far enough to contact the right-hand electrodes. Thus, with centrifugal acceleration less than $0.18g$ there is no compensation.

Consider now a turn in which the centrifugal acceleration exceeds $0.18g$. The mercury in the bank levelling switch is displaced to the end of the tube and so disconnects the normal supply to the bank torque motor, i.e. it now acts as an erection cut-out. However, the pitch levelling switch still responds to a forward tilt and remains connected to its torque motor, and as will be noted from the diagram, it also connects a supply to the lower of the two changeover switches. Since the mercury in these switches is also displaced by the centrifugal acceleration, a circuit is completed from the lower switch to the bank torque motor, which precesses the gyro axis to the right to reduce the bank error. At the same time, the pitch levelling switch completes a circuit to the pitch torque motor, which then precesses the gyro axis rearward so reducing the pitch error. Thus, during turns a constant control is applied about both the pitch and roll axes by the pitch levelling switch; hence the term "pitch-bank erection."

The changeover function of the curved mercury switches depends on the direction of tilt of the gyro axis in pitch. This is indicated by the broken arrows in Fig 5.20, the gyroscope and the pitch levelling switch now being tilted rearward, the latter connects a supply to the upper of the two changeover switches and changes its direction to the bank torque motor causing it to precess the gyroscope to the left.

The change in direction of the supply to the bank torque motor is also dependent on the direction of the turn, as a study of Fig 5.20 will show.

As in the erection cut-out method of compensation, this system requires a "fast-erection" facility to bring the gyro axis to the true vertical when it is tilted more than 10° in either bank or pitch.

Since the forces and torques acting on the gyroscope depend on the aircraft's speed and rate of turn, then obviously all erection errors will vary accordingly, and this makes it rather difficult to provide compensation which will eliminate them entirely. It is usual, therefore, particularly for instruments employing the inclined axis and bank cut-out method of compensation, to base compensation on a standard rate one turn of 180° per minute at an airspeed of 200 m.p.h. At other rates of turn and airspeeds the errors are small.

Pendulosity Errors

Pendulosity, or "bottom heaviness" as it is sometimes called, is often deliberately introduced in gyro horizons so that the gyroscope will always be resting near its vertical position. This helps to reduce the erection time when starting, and also it prevents the gimbal system from spinning about the bank and pitch axes during run-down of the rotor. However, it can be acted upon by accelerating and decelerating forces in straight and level flight, and centrifugal forces during turns; consequently, it is an additional source of error, i.e. pendulosity error.

When acceleration takes place the base of the rotor assembly tends to lag behind owing to inertia, i.e. it tends to swing directly rearwards. In following the force through with the aid of the 90° precession rule, it will be seen, however, that the rotor assembly will precess about the bank axis to port or starboard depending on the direction of rotor rotation. A deceleration has the opposite effect.

The pendulosity error resulting from a turn may be analysed in a manner similar to that of erection errors. In Fig 5.21 an aircraft is again considered as turning to starboard through 360° from the point A. As the turn is entered the

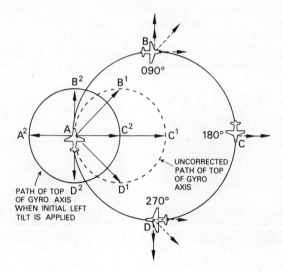

Fig 5.21 Pendulosity error

centrifugal acceleration tends to swing the base of the rotor assembly to port, causing precession of the gyro about the pitch axis, which again depends on the rotation of the rotor. In this instance, the gyro axis tilts forward to a false vertical and the instrument indicates an apparent climb. Throughout the turn, the top of the gyro axis traces out a circular path which, unlike that resulting from turning effects on erection systems, is synchronized with the aircraft's heading change. As before, any chord of the circle from the point at which the turn commenced indicates the tilt of the gyro axis in relation to the true vertical, and varying errors in bank and pitch indications will be presented.

Compensation for Pendulosity Errors

Compensation is usually effected by adopting the inclined-axis method, the inclination in this case being about the bank axis, and the direction being dependent on that of rotor rotation. The amount of inclination is governed by the type of instrument, two typical values being 0·5° and 1·75°.

The effect of the compensation, shown by the full circle in Fig 5.21, is exactly the same as that produced by inclining the gyro axis in pitch, i.e. the top of the axis traces out a circular path about itself to produce a single constant error.

TURN-AND-BANK INDICATORS

The turn-and-bank indicator was the first of the aircraft flight instruments to use a gyroscope as a detecting element, and in conjunction with a magnetic compass, it made a valuable contribution to the art of flying without external references. It was thus considered an essential primary "blind flying" instrument for all types of aircraft. However, with aircraft development, changes in operational requirements, and the introduction of advanced flight instruments and systems, there has been much discussion as to the place a turn-and-bank indicator should occupy in the flight instrument group. In the smaller types of aircraft, it still functions as a primary instrument, but for the larger and more complex aircraft, particularly civil types, it is used mainly to supplement other methods for determining that a turn is being executed at the precise rate demanded by procedural requirements.

Fig 5.22 Typical dial presentations of turn-and-bank indicators

A turn-and-bank indicator contains two independent mechanisms: a gyroscopically controlled pointer mechanism for the detection and indication of the rate at which the aircraft turns, and a mechanism for the detection and indication of bank and/or slip. The dial presentations of two typical indicators are shown in Fig 5.22.

For the detection of rates of turn, direct use is made of gyroscopic precession, and in order to do this the gyroscope is arranged in the manner shown in Fig 5.23. Such an arrangement is known as a *rate gyroscope*.

It will be noted that the gyroscope differs in two respects from those employed in directional gyros and gyro horizons; it has only one gimbal ring and it has a spring connected between the gimbal ring and casing to restrain movement about the fore-and-aft axis. Let us examine Fig 5.23 a little more closely in order to understand how such an arrangement can be made to indicate rates of turn.

When the instrument is in its normal operating position, due to the spring restraint the rotor spin axis will always be horizontal and the turn pointer will be at the zero datum mark. With the rotor spinning, its rigidity will further ensure that the zero condition is maintained.

Let us assume for a moment that the gyroscope has no spring restraint and that the instrument is turned to the left. The gimbal ring being pivoted on the fore-and-aft axis will also turn, but as the rigidity of the gyroscope resists this turning movement it will precess. The direction of precession may be determined by the simple rule already given. A turn to the left causes a force to be applied at the front pivot of the gimbal ring, and this is the same as trying to push the rotor round at the point F on its rim. In following this through 90° in the direction of rotation, precession will take place at point P, thus causing the gimbal ring and rotor to tilt about the fore-and-aft axis. If a pointer were fixed to the gimbal ring, it too would tilt through the same angle and would indicate a turn and also its direction. However, we are more interested in the rate at which a turn is being executed, and to obtain an indication of this, we control the angular deflection of the gimbal ring by connecting it to the instrument case through the medium of a spring.

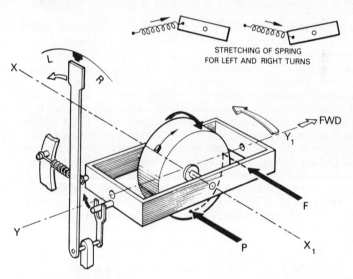

Fig 5.23   Rate gyroscope

Considering once again the left turn indicated, the gyroscope will now precess and will stretch the spring until the force it exerts prevents further deflection of the gyro. Since precession of this type of gyroscope is equal to the product of angular momentum of the gyroscope and the rate of turn, then the spring force is a measure of the rate of turn. If the spring is linear, i.e. its force is proportional to the gimbal ring deflection, and the deflection is small, the actual movement of the gimbal ring from the zero or rest position can be taken as the required measure of turn rate.

In practice the gimbal ring deflection is generally not more than 6°, the reason for this being to reduce the error due to the rate-of-turn component not being at right angles to the spin axis during gimbal ring deflection.

The rate-of-turn pointer is actuated by the gimbal ring and a magnifying system the design of which varies between manufacturers. Scales are calibrated in what are termed "standard" rates, and although not always marked on a scale they are classified by the numbers 1 to 4 and correspond to turn rates of

180, 360, 540 and 720° per minute respectively. The marks shown at either side of zero of the scale in the right-hand upper corner of Fig 5.22 correspond to a Rate 1 turn.

A system for damping out oscillations of the gyroscope is also incorporated and is adjusted so that the turn pointer will respond to fast rate-of-turn changes and at the same time respond to a definite turn rate instantly.

It should be noted that a rate gyroscope requires no erecting device or correction for random precession, for the simple reason that it is always centred by the control spring system. For this reason also, it is unnecessary for the rotor to turn at high speed, a typical speed range being 4,000–4,500 rev/min. The most important factor in connection with speed is that it must be maintained constant, since precession of the rotor is directly proportional to its speed.

Bank Indication

In addition to the primary indication of turn rate, it is also necessary to have an indication that the aircraft is correctly banked for the particular turn. A secondary indicating mechanism is therefore provided which depends for its operation on the effects of gravitational and centrifugal forces. Two principal mechanical methods may be employed: one utilizing a gravity weight and pointer, and the other, a ball in a curved liquid-filled glass tube (see Fig 5.22).

The gravity-weight method is illustrated schematically in Fig 5.24. In normal flight, diagram (*a*), gravity holds the weight in such a position that the pointer indicates zero. At (*b*) the aircraft is shown turning to the left at a certain airspeed and bank angle. The indicator case and scale move with the aircraft, of course, and because of the turn, centrifugal force in addition to that of gravity acts upon the weight and tends to displace it outwards from the centre of the turn. However, when the turn is executed at the correct bank angle then there is a balanced condition between the two forces and so the weight and pointer still remain at the zero position, but this time along the resultant of the two forces. If the airspeed were to be increased during the turn, then the bank angle and centrifugal force would also be increased, but so long as the bank angle is correct the weight and pointer will still remain at the zero position along the new resultant of forces.

If the bank angle for a particular turn rate is not correct, say underbanked as in diagram (*c*), then the aircraft will tend to skid out of the turn. Centrifugal force will predominate under such conditions and displace the weight, and the pointer from its zero position. When the turn is overbanked, as at (*d*), the aircraft will tend to slip into the turn and so the force due to gravity will now have the predominant effect on the weight. The pointer will thus be displaced from zero in the opposite direction to that of an underbanked turn.

The effects of correctly and incorrectly banked turns on the ball-type indicating element are similar to those described in the foregoing paragraphs. The major differences are that the directions of ball displacement are opposite to those of the pointer-type element because the forces act directly on the ball. This is made clear by the series of diagrams in Fig 5.25.

Typical Indicators

The mechanism of a typical air-driven indicator is shown in Fig 5.26. Air enters the instrument through a filter situated at the rear of the case and is led to a jet block via the inlet connecting tube. The jet is set at an angle so that the air is

Primary Flight Instruments (Attitude Indication) 145

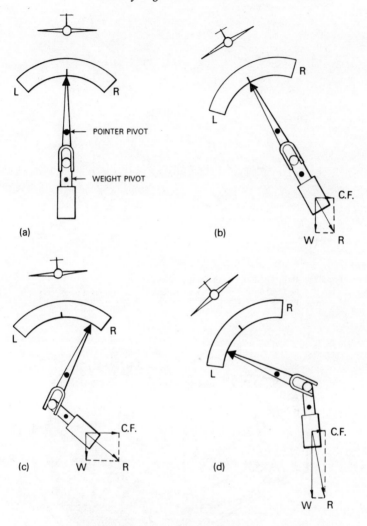

Fig 5.24 Gravity-weight method of bank indication
(a) Level flight
(b) Correctly banked
(c) Underbanked (skidding out of the turn
(d) Overbanked (slipping into the turn)

directed on to the rotor buckets. The direction of rotation is such that, with the indicator installed, a point at the top of the rotor moves in the direction of flight. The pointer moves over the scale, which has a centre-zero mark and a mark at each end: Adjustment of the gyroscope sensitivity is provided by a screw attached to one end of the rate-control spring, the screw protruding through a bracket mounted on the front plate of the mechanism.

A stop is provided to limit the movement of the gimbal ring to an angle which causes slightly more than full-scale deflection (left or right) of the pointer.

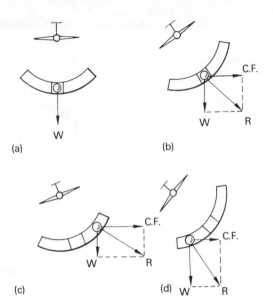

Fig 5.25 Ball-type bank indicating element
(a) Level flight
(b) Correctly banked
(c) Underbanked (skidding out of turn)
(d) Overbanked (slipping into the turn)

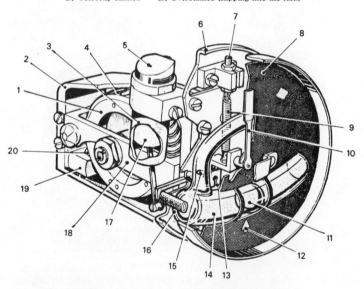

Fig 5.26 Mechanism of an air-driven turn-and-bank indicator

1 Rotor
2 Instrument frame
3 Damping cylinder
4 Buckets
5 Air bleed
6 Front plate
7 Rate-spring adjusting screw
8 Dial
9 Rate spring
10 Pointer
11 Agate ball
12 Datum arrow
13 Gimbal front pivot
14 Slip indicator
15 Expansion chamber
16 Fluorescent card
17 Piston
18 Gimbal ring
19 Jet block
20 Jet

A feature common to all indicators is damping of gimbal ring movement to provide "dead beat" indications. In this particular type, the damping device is in the form of a piston, linked to the gimbal ring, and moving in a cylinder or dashpot. As the piston moves in the cylinder, air passes through a small bleed hole the size of which can be adjusted to provide the required degree of damping.

The slip indicator is of the ball and liquid-filled tube type, the tube and its expansion chamber being concealed behind the dial and clipped in position against a card treated with fluorescent paint. The liquid used is white spirit.

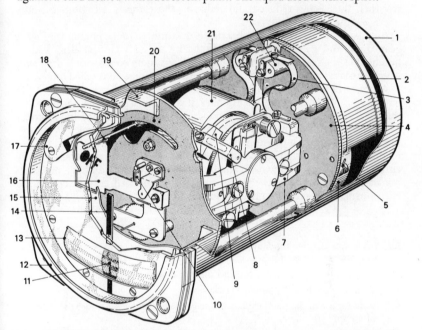

Fig. 5.27 Mechanism of a typical direct-current operated turn-and-bank indicator

1	Case	12	Bezel
2	Suppressor assembly	13	Slip indicator
3	Feed spring	14	Pointer
4	Rear end plate	15	Dial
5	Insulated connector	16	"Off" flag
6	Magnetic damping unit	17	Rate scale
7	Gimbal	18	Stirrup arm
8	Stirrup	19	Dial frame
9	Stirrup magnet	20	Front frame plate
10	Flag spring	21	Gyroscope rotor
11	Ball	22	Brush feed insulator

The mechanism of a typical direct-current operated turn-and-bank indicator is illustrated in Fig 5.27. The gimbal system follows the general pattern adopted for rate gyroscopes, varying only in construction attendant upon electrical operation.

The rotor consists of a lap-wound armature and an outer rim arranged concentrically, the purpose of the outer rim being to increase the rotor mass and radius of gyration. The armature rotates inside a cylindrical two-pole permanent-magnet stator secured to the gimbal ring.

Direct current is fed to the brushes and commutator via a radio-interference suppressor and flexible springs which permit movement of the inner ring. The rotor speed is controlled by two identical symmetrically opposed centrifugal cut-outs. Each cut-out consists of a pair of platinum-tipped governor contacts, one fixed and one movable, which are normally held closed by a governor adjusting spring. Each cut-out has a resistor across its contacts, which are in series with half of the rotor winding. When the maximum rotor speed is attained, centrifugal force acting on the contacts overcomes the spring restraint causing the contacts to open. The armature current therefore passes through the resistors, thus being reduced and reducing the rotor speed. Both cut-outs operate at the same critical speed.

Angular movement of the gimbal ring is transmitted to the pointer through a gear train, and damping is accomplished by an eddy-current drag system mounted at the rear of the gyro assembly. The system consists of a drag cup, which is rotated by the gimbal ring, between a field magnet and a field ring.

A power-failure warning flag is actuated by a stirrup arm pivoted on the gimbal ring. When the rotor is stationary, the stirrup arm is drawn forward by the attraction between a magnet mounted on it and an extension (flux diverter) of the permanent-magnet stator. In this condition the flag, which is spring-loaded in the retracted position, is depressed by the stirrup arm so that the OFF reading appears through an aperture in the dial. As rotor speed increases, eddy currents are induced in the rotor rim by the stirrup magnet, and at a predetermined speed, reaction between the magnet and induced current causes the stirrup arm to lift and the OFF reading to disappear from view.

QUESTIONS

5.1 (a) Define the two fundamental properties of a gyroscope.
(b On what factors do these properties depend?

5.2 How are the gyroscopic properties utilized in flight instruments?

5.3 What is meant by "apparent" precession?

5.4 With the aid of diagrams explain how a gyroscope precesses under the influence of an applied torque.

5.5 Discuss the basic principles and usage of (a) a space gyroscope, (b) a "tied" gyroscope, (c) an "earth" gyroscope, (d) a rate gyroscope. (SLAET)

5.6 What methods are adopted for driving the rotors of gyroscopic flight instruments?

5.7 How is the gyroscopic principle applied to a gyro horizon?

5.8 Describe the construction and operation of an electrically-driven gyro horizon including any special design features. (SLAET)

5.9 How are the gyroscopes of gyro horizons erected to and maintained in their normal operating position?

5.10 Describe the construction and operation of the erection system of a gyro horizon with which you are familiar.

5.11 Explain how the magnetic field set up in the stator of a torque motor is made to rotate.

5.12 (a) What are the functions of a "fast-erection" system?
(b) With the aid of a circuit diagram explain the operation of the levelling switch method.

5.13 What precautions must be taken when using the levelling switch method of fast erection?

5.14 (a) What effects does acceleration of an aircraft have on the indications of a gyro horizon?
(b) What do you understand by the terms "erection error" and "pendulosity error"?

5.15 What methods are adopted for the compensation of "erection error"? Describe the operation of a method with which you are familiar.

5.16 How is compensation for "pendulosity error" usually effected?

5.17 Describe how the rate gyroscope principle is applied to a turn-and-bank indicator.

5.18 (a) Why is it unnecessary to incorporate an additional erecting device in a turn-and-bank indicator?
(b) Why is it important for the gyro of this indicator to rotate at a constant speed?
(c) Describe how a constant speed is maintained in a d.c.-operated instrument.

5.19 With the aid of diagrams, describe how a ball type of bank indicator indicates (a) a correctly banked turn, (b) a turn to starboard in which the aircraft is overbanked.

6 Primary Heading Indicating Instruments

DIRECT-READING MAGNETIC COMPASSES

Direct-reading magnetic compasses were the first of the many airborne flight and navigational aids ever to be introduced in aircraft. Their primary function is to show the direction in which an aircraft is heading with respect to the earth's magnetic meridian.

As far as present-day aircraft and navigational aids are concerned, however, a subdivision of this function has been brought about by the type of aircraft and by the aids employed. For example, in many small aircraft the magnetic compass is utilized as the primary heading indicator, while in aircraft employing remote-indicating compasses and other advanced navigational aids, it plays the role of standby heading indicator.

The principle on which they operate is the very basic one of reaction between the magnetic field of a suitably suspended permanent magnet and the field surrounding the earth. This principle and certain other fundamentals of magnetism will no doubt have already been studied by readers, but nevertheless the reminders of this and the next few pages should prove useful.

Properties (Fig 6.1)

First of all let us consider the three principal properties of a permanent magnet: (i) it will attract other pieces of iron and steel, (ii) its power of attraction is concentrated at each end, and (iii) when suspended so as to move horizontally, it always comes to rest in an approximately North–South direction. The second and third properties are related to what are termed the *poles* of a magnet, the end of the magnet which seeks North being called the *North pole* and the end which seeks South the *South pole*.

When two such magnets are brought together so that both North poles or both South poles face each other, a force is created which keeps the magnets apart. When either of the magnets is turned round so that a North pole faces a South pole again a force is created, but this time to pull the magnets more closely to each other. Thus, *like poles repel and unlike poles attract*; this is one of the fundamental laws of magnetism. The force of attraction or repulsion between two poles *varies inversely as the square of the distance between them*.

The region in which the force exerted by a magnet can be detected is known as a *magnetic field*. Such a field contains *magnetic flux*, which can be represented in direction and density by *lines of flux*. The conventional direction of the lines of flux outside a magnet is from the North pole to the South pole. The lines are continuous and unbroken, so that inside the magnet their direction is from South pole to North pole. If two magnetic fields are brought close together, their lines of flux do not cross one another but form a distorted pattern, still consisting of closed loops.

The symbol for magnetic flux is Φ, and its unit is the *weber* (Wb). The amount

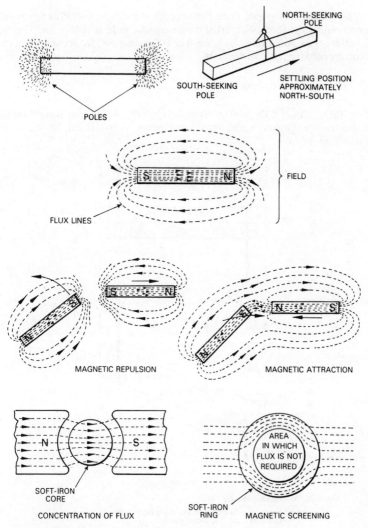

Fig 6.1 Fundamental magnetic properties

of flux through unit area, indicated by the spacing of the lines of flux, is known as *magnetic flux density* (B); its unit is the weber per square metre, or *tesla* (T).

Magnetic flux is established more easily in some materials than in others: in particular it is established more easily in magnetic materials than in air. All materials, whether magnetic or not, have a property called *reluctance* which resists the establishment of magnetic flux and is equivalent to the resistance of an electric circuit. It follows that, if a material of low reluctance is placed in a magnetic field, the flux density in the material will be greater than that in the surrounding air.

152 Aircraft Instruments

Magnetic field strength, H, or the strength of a magnetic field at any point is measured by the force, F, exerted on a magnetic pole at that point. The force depends on the *pole strength*, i.e. the flux Φ "emanating" from the pole* as well as on the field strength. In symbols,

$$H = \frac{F}{\Phi} \quad \text{newtons per weber}$$

Thus the unit of H is the *newton per weber* (N/Wb). A unit that is more familiar to electrical engineers is the *ampere per metre* (A/m). It can be shown that 1 N/Wb = 1 A/m.

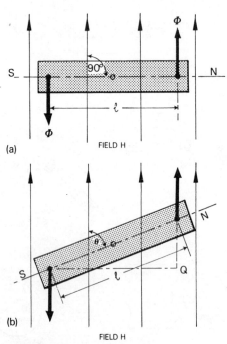

Fig 6.2 Magnetic moment
(a) Magnet at right angles to a uniform field
(b) Magnet at angle θ to a uniform field

Magnetic Moment

The *magnetic moment* of a magnet is the tendency for it to turn or be turned by another magnet. It is a requirement in aircraft compass design that the strength of this moment be such that the magnetic detecting system will quickly respond to the directive force of a magnetic field, and in calculating it the length and pole strength of a magnet must be considered.

In Fig 6.2, suppose the pivoted magnet shown at (a) is of pole strength Φ and the length of its magnetic axis is l, then its magnetic moment m is equal to the product of the pole strength and magnetic length, thus: $m = l\Phi$.

* "Emanating" from the pole if it is a North pole; "returning" to it if it is a South pole.

If now the magnet is positioned at right angles to a uniform magnetic field H, the field will be distorted in order to "pass through" the magnet. In resisting this distortion, the field will try to pull the magnet into alignment with it. Each pole will experience a force of ΦH newtons, and as the forces act in opposite directions they constitute a couple. Now, the torque, M, of a couple is the product of one of the equal forces and the perpendicular distance between them, i.e. $M = l\Phi H$; but $l\Phi = m$, so that $M = mH$.

From the foregoing it is thus evident that the greater the pole strength and the longer the magnet, the greater will be its tendency to turn into line with a surrounding magnetic field. Conversely, the greater will be the force it exerts upon the surrounding field, or indeed upon any magnetic material in its vicinity.

In Fig 6.2 (b), the magnet needle is shown inclined at an angle θ to the field H. The force on each pole is still ΦH, but the perpendicular distance between the forces is now SQ. Now $SQ/SN = SQ/l = \sin \theta$; therefore $SQ = l \sin \theta$. Thus the torque acting on the magnet at an angle θ is $l\Phi H \sin \theta$, or $mH \sin \theta$.

Magnet in a Deflecting Field

In Fig 6.3, a magnet is situated in a uniform magnetic field H_1 and a uniform deflecting field H_2 is applied at right angles to H_1. When the magnet is at an angle θ to field H_1, as already shown, the torque due to H_1 is $mH_1 \sin \theta$. The

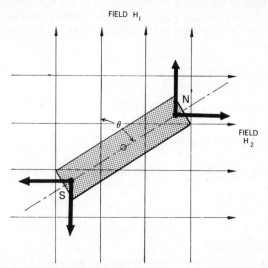

Fig 6.3 Magnet in a deflecting field

torque due to H_2 is $mH_2 \cos \theta$. Thus, for the magnet to be in equilibrium, i.e. subjected to equal and opposite torques, $mH_1 \sin \theta = mH_2 \cos \theta$, so that the strength of the deflecting field is $H_2 = H_1 \tan \theta$.

Period of a Suspended Magnet

If a suspended magnet is deflected from its position of rest in the magnetic field under whose influence it is acting, it at once experiences a couple urging it back into that position, and when the deflecting influence is removed the magnet, if undamped, will oscillate backwards and forwards about its equilibrium position before finally coming to rest. The time taken for the magnet to swing from one extremity to another and back again, i.e. the time for a complete vibration, is known as the *period* of the magnet.

As the magnet gradually comes to rest, the amplitude of the vibration gradually gets less but the period remains the same and it cannot be altered by adjusting the amplitude. The period of a magnet depends upon its shape, size or mass (factors which affect the moment of inertia), its magnetic moment, and the strength of the field in which it vibrates. The period varies with these factors in the following ways: (i) it grows longer as the mass increases; (ii) it becomes shorter as the field strength increases.

The vibrations of a magnet acting under the influence of a magnetic field are very similar to those of an ordinary pendulum swinging under the influence of gravity; the period T of a pendulum is given by

$$T = 2\pi \sqrt{\frac{l}{g}}$$

where l is the length of the pendulum and g the acceleration due to gravity.

When a magnet of magnetic moment m is displaced through an angle θ then, as already shown, the torque, T, restoring it to the equilibrium position is $mH \sin \theta$. If I is the moment of inertia* of the vibrating magnet about an axis through its centre of gravity perpendicular to its length, then its angular acceleration is

$$\alpha = \frac{M}{I} = \frac{mH \sin \theta}{I}$$

If the displacement is small, $\sin \theta$ and θ do not differ appreciably, so that

$$\frac{mH \sin \theta}{I} \text{ may be written } \frac{mH}{I}$$

and is constant. The motion is simple harmonic, having a period given by

$$T = 2\pi \sqrt{\frac{I}{mH}}$$

Hard Iron and Soft Iron

"Hard" and "soft" are terms used to qualify varieties of magnetic materials according to the ease with which they can be magnetized. Metals such as cobalt and tungsten steels are of the hard type since they are difficult to magnetize but once in the magnetized state they retain the property for a considerable length of time; hence the term *permanent magnetism*. Metals which are easy to magnetize (silicon-iron for example), and generally lose their magnetic state once the magnetizing force is removed, are classified as soft.

These terms are also used to classify the magnetic effects occurring in aircraft, a subject which is dealt with in detail in Chapter 8.

---

* Other symbols for moment of inertia are K and J.

TERRESTIAL MAGNETISM

The surface of the earth is surrounded by a weak magnetic field which culminates in two internal *magnetic poles*, situated near the North and South *true* or *geographic poles*. That this is so is obvious from the fact that a magnet freely suspended at various parts of the earth's surface will be found to settle in a definite direction, which varies with locality. A plane passing through the magnet and the centre of the earth would trace on the earth's surface an imaginary line called the *magnetic meridian* as shown in Fig 6.4.

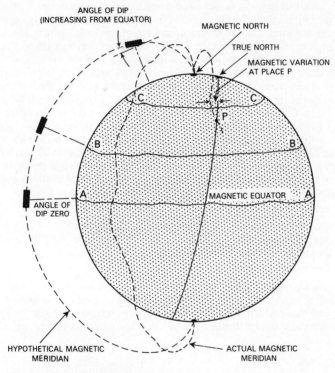

Fig 6.4 Terrestial magnetism
   Lines AA, BB and CC are isoclinals

It would thus appear that the earth's magnetic field is similar to that which would be expected at the surface if a short but strongly magnetized bar magnet were located at the centre. This partly explains the fact that the magnetic poles are relatively large areas, due to the spreading out of the lines of force and it also gives a reason for the direction of the field being horizontal in the vicinity of the equator. However, the origin of the field is still not exactly known, but for purposes of explanation, the supposition of a bar magnet at the earth's centre is useful in visualizing the general form of the magnetic field as it is known to be.

The earth's magnetic field differs from that of an ordinary magnet in several respects. Its points of maximum intensity, or strength, are not at the magnetic poles (theoretically they should be) but occur at four other positions, two near each pole, known as *magnetic foci*. Moreover, the poles themselves are continually changing their positions, and at any point on the earth's surface the field is not symmetrical and is subject to changes both periodic and irregular.

Magnetic Variation

As meridians and parallels are constructed with reference to the geographic North and South poles, so can magnetic meridians and parallels be constructed with reference to the magnetic poles. If a map were prepared to show both true and magnetic meridians, it would be observed that these intersect each other at angles varying from 0° to 180° at different parts of the earth, diverging from each other sometimes in one direction and sometimes in the other. The horizontal angle contained between the true and magnetic meridian at any place is known as the *magnetic variation* or *declination*.

When the direction of the magnetic meridian inclines to the left of the true meridian at any place, the variation is said to be westerly. When the inclination is to the right of the true meridian the variation is said to be easterly. It varies in amount from 0° along those lines where the magnetic and true meridians run together to 180° in places between the true and magnetic poles. At some places on the earth where the ferrous nature of the rock disturbs the earth's main magnetic field, local attraction exists and abnormal variation occurs which may cause large changes in its value over very short distances. While the variation differs all over the world, it does not maintain a constant value in any one place, and the following changes, themselves not constant, may be experienced:

(i) *Secular change*, which takes place over long periods due to the changing positions of the magnetic poles relative to the true poles.
(ii) *Annual change*, which is a small seasonal fluctuation superimposed on the secular change.
(iii) *Diurnal change* (daily).

Information regarding magnetic variation and its changes is given on special charts of the world which are issued every few years. Lines are drawn on the charts, and those which join places having equal variation are called *isogonal lines*, while those drawn through places where the variation is zero are called *agonic lines*.

Magnetic Dip

As stated earlier, a freely suspended magnet needle will settle in a definite direction at any point on the earth's surface and will lie parallel to the magnetic meridian at that point. However, it will not lie parallel to the earth's surface at all points for the reason that the lines of force themselves are not horizontal as may be seen from Fig 6.4. These lines emerge vertically from the North magnetic pole, bend over and descend vertically into the South magnetic pole, and it is only at what is known as the *magnetic equator* that they pass horizontally along the earth's surface. If, therefore, a magnetic needle is carried along a meridian from North to South, it will be on end, red end down, at the start, horizontal near the equator and finish up again on end but with the blue end down.

The angle the lines of force make with the earth's surface at any given place is called the *angle of dip* or *magnetic inclination*, and varies from 0° at the magnetic equator to 90° at the magnetic poles. Dip is conventionally considered positive when the red end of a freely suspended magnet needle dips below the horizontal, and negative when the blue end dips below the horizontal. Hence all angles of dip north of the magnetic equator will be positive, and all angles of dip south of the magnetic equator will be negative.

The angle of dip at all places undergoes changes similar to those described for variation and is also shown on charts of the world. Places on these charts having the same magnetic dip are joined by lines known as *isoclinals*, while those at which the angle is zero are joined by a line known as the *aclinic line* or *magnetic equator*, of which mention has already been made.

Earth's Total Force or Magnetic Intensity

When a magnet needle freely suspended in the earth's field comes to rest, it does so under the influence of the total force of the earth's magnetism. The value of this total force at a given place is difficult to measure, but seldom needs to be known. It is usual, therefore, to resolve this total force into its horizontal and vertical components, termed H and Z respectively; if the angle of dip θ is known, the total force can be calculated.

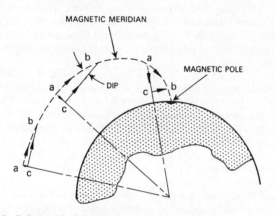

Fig 6.5 Relationship between dip, Z, H and total force

$a - c$ = Vertical component Z
$c - b$ = Horizontal component H
$a - b$ = Total force T

Given angle of dip θ and H,

$\dfrac{Z}{H} = \tan \theta$ and $Z = H \tan \theta$

$\dfrac{H}{T} = \cos \theta$ and $T = H \cos \theta$

$T^2 = H^2 + Z^2$

A knowledge of the values of the horizontal and vertical components is of great practical value, particularly in connection with compass deviation and adjustment. Both components are responsible for the magnetization of any magnetic parts of the aircraft which lie in their respective planes, and may therefore fluctuate at any place for different aircraft or for different compass positions in the same aircraft. The relationship between dip, horizontal force, vertical force and total force is shown in Fig 6.5.

As in the case of variation and dip, charts of the world are published showing the values of H and Z for all places on the earth's surface, together with the mean annual change. Lines of equal horizontal and vertical force are referred to as *isodynamic lines*.

The earth's magnetic force may be stated either as a relative value or an absolute value. If stated as a relative value and in connection with aircraft compasses this is the case, it is given relative to the horizontal force at Greenwich.

TYPES OF COMPASS

Although aircraft compasses may differ in their physical dimensions and methods of presenting heading information, we can say that the types now in use have the following common principal features: a magnet system housed in a bowl; liquid damping; and liquid expansion compensation.

Magnet Systems

Two typical magnet systems are illustrated in Fig 6.6. The one at (*a*) is made up of four small cylindrical permanent magnets supported in vertical parallel pairs in a wire frame forming part of the indicating element. The reason for using a number of magnets instead of a single magnet is to provide a large magnetic moment and at the same time to keep the moment of inertia as small as possible, thus helping to prevent overswinging of the element in responding to the earth's magnetic field.

In addition to the supporting frame for the magnets, damping wires are affixed to the indicating element together with two wire filaments at 90° to each other. One of these wires is positioned parallel to the magnets and has one end in the form of a cross to indicate the North-seeking ends. Short lengths of minute-bore glass capillary tube filled with luminous or fluorescent compound are cemented to the wires to facilitate observation at night.

The system shown at (*b*) is a much simpler arrangement, being designed for a smaller type of card compass and consisting solely of a single annular cobalt-steel magnet.

The method of suspending a magnet system is more or less the same for all types of compass and consists of an iridium-tipped pivot secured to the centre of the magnet system and resting in a sapphire cup supported in a holder or stem. The use of iridium and sapphire in combination provides hard wearing properties and reduces pivot friction to a minimum.

A compass bowl may be constructed of brass or of moulded plastic (Diakon) depending on the compass type. The interior of metal bowls is coated with a special paint which is resistant to the effects of the liquid used for magnet system damping. A lubber mark, or line against which headings are observed, is fixed to the interior of the bowl and lies on or parallel to the fore-and-aft line when the compass is installed in an aircraft.

Liquid Damping

The primary reason for filling compass bowls with a liquid is to make the compass *aperiodic*. This is a term we apply to a compass whose magnet system, after being deflected, will return to its equilibrium position directly without oscillating or overshooting. Other reasons for using a liquid are that it steadies the magnet system and gives it a certain buoyancy, thereby reducing the weight on the pivot and so diminishing the effects of friction and wear.

The liquids, which may be of the mineral or alcohol type, must meet such requirements as low freezing point, low viscosity, high resistance to corrosion, and freedom from discoloration.

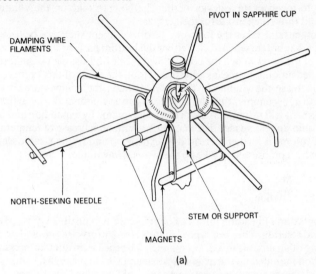

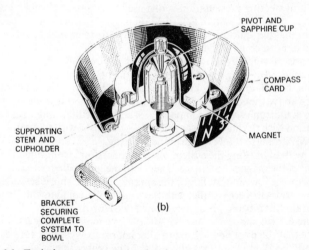

Fig 6.6 Typical compass magnet systems
 (a) Filament type   (b) Circular magnet and card

Liquid Expansion Compensation

Compass liquids are subject to expansion and contraction with changes in temperature, and the resulting changes in their volume can have undesirable effects. For example, with reduction of temperature the liquid would contract and so leave an air space in the bowl thus reducing the damping effect. Conversely, expansion would take place under high temperature conditions tending to force the liquid out and resulting in leaks around bowl seals. It is therefore necessary to incorporate a device within the bowl to take up the volumetric changes and thus compensate for their effects.

The compensator takes the form of a flexible element such as a bellows or a corrugated diaphragm which forms either the bottom part, or in some cases the rear part, of the bowl. When the bowl is filled the flexible element is compressed by a specified amount by means of a special tool, the effect of this compression being to increase the volume of the bowl. If now, the compensated bowl is subjected to a low temperature the liquid will contract, but at the same time the flexible element will respond to the decrease in volume by expanding and filling up with liquid any air spaces that may form. With an increase of temperature, the liquid volume is further increased by expansion and so the flexible element will be further compressed to take up the increase in volume.

Heading Presentation

The presentation of heading information falls into two distinct types: (i) card, and (ii) grid steering. The card type is quite a straightforward one and presents the heading of an aircraft in a direct manner. The card is an annular pressing of conic section mounted on the magnet system and visible through a window of a metal-bowl type compass or a transparent part of a moulded-plastic bowl type.

The grid steering presentation is one used in compasses employing the filament-type of magnet system described on page 158. It was developed to assist pilots in making quicker settings of a course to steer and to instantly show whether or not a course is being maintained.

It consists essentially of two main parts: the wire filament, which is attached to the magnet system and indicates alignment of the magnet system with magnetic North, and a ring known as the grid ring which can be rotated and clamped above the compass bowl as required. The grid ring itself is graduated in degrees in the same way as a compass card, and to facilitate alignment of its 000° graduation with the filament, the grid ring is provided with two wires which are arranged parallel to the N–S line of the grid ring.

The method of using the system may be understood from Fig 6.7. At (*a*) the filament of the magnet system is aligned with the lubber line, which lies on, or is parallel to, the fore-and-aft line of the aircraft. The grid ring is set to North, and since the grid wires are parallel with the magnet-system filament, they indicate that the aircraft is being accurately steered due North.

Figure 6.7 (*b*) shows the aircraft 10° off its course and to starboard. Let us suppose that the pilot now wishes to alter course to West. He rotates the grid ring until the W graduation is aligned with the lubber line as shown at (*c*); he then turns his aircraft to starboard carrying with it the compass and grid ring.

After turning through 90° the position will be as shown at (*d*). The grid wires and the magnet-system filament are again parallel, but the aircraft is now heading West, as indicated by the alignment of the lubber line with the W graduation on the grid wing.

In order to minimize the risk of getting on to the reciprocal of the course it is intended to steer, the S ends of the grid wires are omitted.

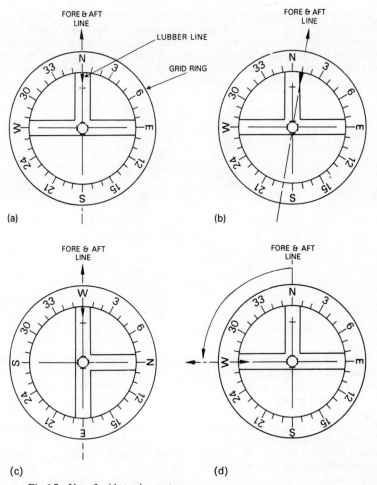

Fig 6.7 Use of grid steering system
(a) Steering due North
(b) 10° off course to starboard
(c) Steering due North, grid ring set to steer West
(d) Steering due West

The Effect of Dip on a Compass Magnet System

Dip, as we already know, is the angle that a suspended magnet needle makes with the horizontal at any particular place owing to the influence of the vertical component Z of the earth's field. Likewise it is known that, for accurate indication of magnetic heading, we are dependent only on the effect of the horizontal component H. This being so, then maximum directional accuracy can only be obtained at places where dip is at, or approaching, zero, and since this is only possible at the equator and its neighbouring latitudes, then a compass utilizing a magnet system uncompensated for the effects of dip, would be very limited in its application as a heading indicator.

A compass must therefore be designed so as to neutralize the effects of the vertical component Z over a much greater range of latitudes, enabling its magnet system to remain horizontal or nearly so. There are several ways of doing this, but the method which has always proved to be the most effective is to make the magnet system pendulous, i.e. to pivot it at a point above the centre of gravity, as shown in Fig 6.8.

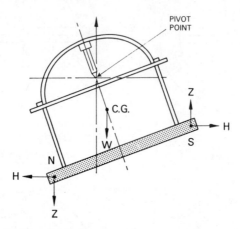

Fig 6.8 Compensation of magnetic dip

When the vertical component Z acts on the magnet system, the latter is caused to tilt, drawing the centre of gravity of the system away from its position below the pivot point. A force is now acting upward through the pivot and a second force is acting downward through the centre of gravity, and as both forces are not acting along the same line, a righting couple is introduced. The couple tends to bring the magnet system once more into the horizontal position. However, the vertical component Z is still being exerted on the magnet system so that it will not return to the horizontal position exactly, except, of course, at the magnetic equator, where the value of Z is zero.

Compasses are designed with a pendulosity such that the magnet system is within approximately 2° of the true horizontal (when the vertical component Z and gravity are the only forces acting on it) between latitudes 60° North and 40° South.

TYPICAL COMPASSES

The card-type compass shown in Figs 6.9 (a) and 6.10 was originally designed for emergency use in military aircraft and to be small and light enough to enable a pilot to carry it in a pocket of his flying overalls. It proved to be quite an accurate heading indicator and has been developed for use as a standard part of present-day aircraft instrumentation.

The magnet system comprises an annular cobalt-steel magnet and a light-alloy card mounted so as to be close to the inner face of the bowl, thereby minimizing errors in observation due to parallax. The card is graduated in

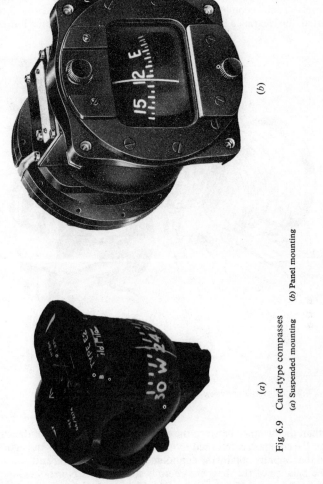

Fig 6.9 Card-type compasses
(a) Suspended mounting   (b) Panel mounting

increments of 10°, intermediate indications being estimated by interpolation. Observations are made against a lubber line moulded on the inner face of the bowl.

Suspension of the system is by means of the usual iridium-tipped pivot and sapphire cup, the latter being supported in a holder mounted on a stem and bracket assembly, secured to the rear of the bowl. To prevent the magnet assembly from becoming detached from the stem, should the compass be inverted, the clearance between the top of the pivot nut and the bowl ceiling is

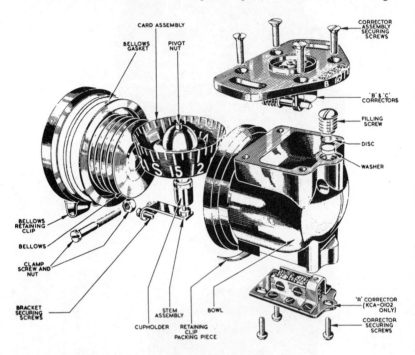

Fig 6.10 Exploded view of typical card compass

less than the distance between the sapphire cup and the top of the cupholder. When the compass is returned to its normal attitude, the pivot is guided back on to the sapphire cup by the cupholder, which has sloping and polished sides.

The balance of the magnet system is such that its North-seeking end is 2° down to compensate for the angle of dip (see page 161).

The bowl is moulded in Diakon and is painted on the exterior with matt black enamel, except for a small area at the front through which the card is observed. This part of the bowl is so moulded that it has a magnifying effect on the card and its graduations. The damping liquid is a silicone fluid (dimethylsiloxane-polymer), $1\frac{1}{2}$ oz being required to fill the bowl. Changes in liquid volume due to temperature changes are compensated by a bellows type of expansion device, secured to the rear of the bowl.

Primary Heading Indicating Instruments   165

The effects of deviation due to fore-and-aft and athwartships components of aircraft magnetism are compensated by permanent-magnet corrector assemblies secured to the compass mounting-plate. An additional corrector assembly may also be provided in some versions of this compass for use in aircraft requiring compensation of a vertical magnetic component.

The card-type compass of Fig 6.9 (b) is designed for direct panel mounting, and employs a needle-type magnet system carrying a card similar to that employed in an E2 compass. The brass case forms the bowl and is sealed by a

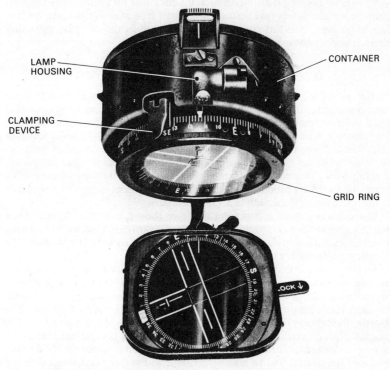

Fig 6.11   Compass employing grid steering system

front bezel plate and a cover at the rear of the bowl. Changes in liquid volume are compensated by a diaphragm type of expansion device sandwiched between the rear part of the bowl and rear cover. A permanent-magnet deviation compensator is mounted on top of the bowl, the adjusting spindles being accessible from the front of the compass. A small lamp is provided for illuminating the card.

A needle type of compass employing the grid steering system is shown in Fig 6.11. It is designed for mounting in the roof of a cockpit at a point above the main instrument panel, its indications being observed with the aid of an adjustable mirror. The bowl is made of brass and is sealed by a glass cover and brass ring known respectively as the verge glass and verge ring; a bellows type of expansion device is integral with the bowl.

As the compass is of the inverted type the stem supporting the magnet system is of necessity secured to the centre of the verge glass. Similarly, it is nesessary to have an additional scale around the outer part of the grid-ring to facilitate setting. The grid-ring, which is also fitted with a glass cover, may be locked on any desired setting by means of a friction-type clamping device.

The bowl and grid-ring assemblies are supported on spring-type anti-vibration mountings inside a container to which are secured three compass fixing lugs. The rearmost lug is graduated to provide coefficient A correction up to $\pm 5°$. A lamp housing is also secured to the container for illuminating the grid-ring outer scale. A separate permanent-magnet type of deviation compensator has to be used with this compass and is secured to the compass mounting bracket directly above the magnet system.

ACCELERATION AND TURNING ERRORS

In the quest for accuracy of an indicating system, it is often found that the methods adopted in counteracting undesirable errors under one set of operating conditions are themselves potential sources of error under other conditions. For example, when a compass magnet system is made pendulous to counteract the effects of dip, the compass can be used over a greater range of latitudes without significant error; but unfortunately, any manoeuvre which introduces a component of aircraft acceleration, either easterly or westerly of the earth's magnetic meridian, produces a torque about the magnet system's vertical axis causing it to rotate in azimuth to a false meridian.

There are two main errors resulting from these acceleration components, namely *acceleration error* and *northerly turning error*, but before considering them in detail it is useful to consider first the effect which would be produced if a simple pendulum were to be suspended in an aircraft.

So long as a constant course and speed are maintained, the pendulum will remain for all practical purposes in the true vertical with its centre of gravity directly below the point of suspension. If, however, the aircraft turns, accelerates or decelerates, the pendulum will cease to be vertical. This is because, owing to inertia, the centre of gravity will lag relative to the pivot and move from the normal position vertically below it. During a correctly banked turn, the forces acting on the centre of gravity will cause the pendulum to remain vertical to the plane of the aircraft and to take up an angle to the true vertical equal to the angle of bank.

Since turns themselves are, in effect, acceleration towards their centres, and whether correctly or incorrectly banked, always cause a pendulum to take up a false vertical, it may be stated broadly that any acceleration or deceleration of the aircraft will cause the centre of gravity of a pendulum to be deflected from its normal position vertically below the point of suspension.

From the foregoing, it is thus apparent that a magnet system suspended to counteract the effects of dip will behave in a similar manner to a pendulum; any acceleration or deceleration in flight results in a displacement of the centre of gravity of the system from its normal position.

Acceleration Error

Acceleration error may be broadly defined as the error, caused by the effect of the vertical component of the earth's field, in the directional properties of a suspended magnet system when the centre of gravity of the system is displaced from its normal position, such errors being governed by the heading on which acceleration or deceleration takes place.

The force applied by an aircraft, when accelerating or decelerating on any fixed heading, is applied to the magnet system at the point of suspension, which is, of course, its only connection. The reaction to this force will be equal and opposite and must act through the centre of gravity below the point of suspension and offset from it due to the slight dip of the magnet system. The two forces constitute a couple which, dependent on the heading being flown by the aircraft, causes the magnet system merely to change its angle of dip or to rotate in azimuth.

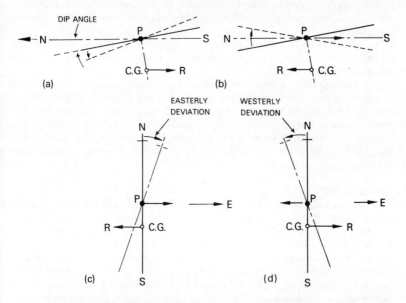

Fig 6.12 Acceleration errors

(a) Acceleration on Northerly heading in Northern Hemisphere
(b) Deceleration on Northerly heading in Northern Hemisphere
(c) Acceleration on Easterly heading in Northern Hemisphere
(d) Deceleration on Easterly heading in Northern Hemisphere

Consider first an aircraft flying in the northern hemisphere and accelerating on a northerly heading. The forces brought into play by the acceleration will be as shown in Fig 6.12 (a). Since both the point of suspension P and the centre of gravity are in the plane of the magnetic meridian, the reaction R causes the N end of the magnet system to nose-down, thus increasing the dip angle without any azimuth rotation. Conversely, when the aircraft decelerates, the reaction at the centre of gravity tilts the needle down at the S end, as shown at (b).

When an aircraft is flying in either the northern or southern hemisphere and changing speed on headings other than the NS meridian, such changes will produce azimuth rotation of the magnet system and errors in indication.

Let us now consider the effects on the compass magnet system when an aircraft flying in the northern hemisphere accelerates on an easterly heading (Fig 6.12 (c)). The accelerating force will again act through the point of suspension P and the reaction R through the centre of gravity, but this time they are acting away from each other at right angles to the plane of the magnetic meridian. The couple will now tend to rotate the magnet system in a clockwise direction, thus indicating an apparent turn to the north, or what is termed easterly deviation. When the aircraft decelerates as at (d) the reverse effect will occur, the couple now tending to turn the magnet system in an anticlockwise direction, indicating an apparent turn to the south, or westerly deviation.

Hence, in the northern hemisphere, acceleration causes easterly deviation on easterly headings, and westerly deviation on westerly headings, whilst deceleration has the reverse effect. In the southern hemisphere the result will be reversed in each case.

As northerly or southerly headings are approached, the magnitude of the apparent deviation decreases, the acceleration error varying as the sine of the compass heading.

One further point may be mentioned in connection with errors brought about by accelerations and decelerations, and that is the effect of aircraft attitude changes. If an aircraft flying level is put into a climb at the same speed, the effect on the compass magnet system will be the same as if the aircraft had decelerated, because the horizontal velocity has changed. If the change in attitude is also accompanied by a change in speed, the apparent deviation may be quite considerable.

Turning Errors

When an aircraft executes a turn, the point of suspension of the magnet system is carried with it along the curved path of the turn, whilst the centre of gravity, being offset, is subjected to the force of centrifugal acceleration produced by the turn, causing the system to swing outwards and to rotate so that apparent deviations, or turning errors, are observed. In addition, during the turn the magnet system tends to maintain a position parallel to the transverse plane of the aircraft thus giving it a lateral tilt the angle of which is governed by the aircraft's bank angle. For a correctly banked turn, the angle of tilt would be maintained equal to the bank angle of the aircraft, because the resultant of centrifugal force and gravity lies normal to the transverse plane of the aircraft and also to the plane through the magnet system's point of suspension and centre of gravity. In this case, centrifugal force itself would have no effect other than to exert a pull on the centre of gravity and so decrease the natural dip angle of the magnet system.

However, as soon as the magnet system is tilted, and regardless of whether or not the aircraft is correctly banked, the system is free to move under the influence of the earth's vertical component Z which will have a component in the lateral plane of the system causing it to rotate and further increase the turning error.

The extent and direction of the turning error is dependent upon the aircraft heading, the angle of tilt of the magnet system, and the dip. In order to form a clearer understanding of its effects on compass direction-indicating properties, we may consider a few examples of aircraft heading changes from the magnetic meridian and in both the northern and southern hemispheres.

Turning from a Northerly Heading towards East or West

Figure 6.13 (*a*) shows the magnet system of a compass in an aircraft flying in the northern hemisphere and on a northerly heading; the north-seeking end of the system is coincident with the lubber line. Let us assume now that the pilot wishes to make a change in heading to the eastward. As soon as the turn commences, the centrifugal acceleration acts on the centre of gravity causing the system to rotate in the same direction as the turn, and since the system is tilted, the earth's vertical component Z exerts a pull on the N end causing further rotation of the system. Now, the magnitude of magnet system rotation is dependent on the rate at which turning and banking of the aircraft is carried out,

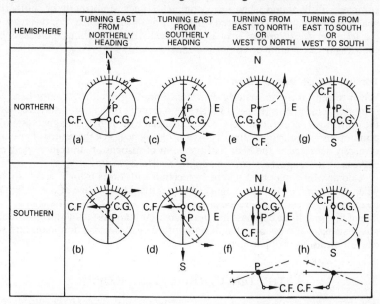

Fig 6.13 Turning errors

and resulting from this three possible indications may be registered by the compass: (i) a turn of the correct sense, but smaller than that actually carried out when the magnet system turns at a slower rate than the aircraft; (ii) no turn at all when the magnet system and the aircraft are turning at the same rate; (iii) a turn in the opposite sense when the magnet system turns at a rate faster than the aircraft. The same effects will occur if the heading changes from N to W whilst flying in the northern hemisphere.

In the southern hemisphere (Fig 6.13 (*b*)) the effects are somewhat different. The south magnetic pole is now the dominant pole and so the natural dip angle changes to displace the centre of gravity to the north of the point of suspension.

We may again consider the case of an aircraft turning from a northerly heading to the eastward. Since the centre of gravity is now north of the point of suspension, the centrifugal acceleration acting on it causes the magnet system to rotate more rapidly in the opposite direction to the turn, i.e. indicating a turn in the correct sense but of greater magnitude than is actually carried out.

Turning from a Southerly Heading towards East or West

If the turns are executed in the northern hemisphere (Fig 6.13 (c)) then because the magnet system's centre of gravity is still south of the point of suspension, the rotation of the system and the indications registered by the compass will be the same as when turning from a northerly heading in the northern hemisphere.

In turning from a southerly heading in the southern hemisphere (Fig 6.13 (d)) the magnet system's centre of gravity is north of the point of suspension and produces the same effects as turning from a northerly heading in the southern hemisphere.

In all the above cases, the greatest effect on the indicating properties of the compass will be found when turns commence near to northerly or southerly headings, being most pronounced when turning through north. For this reason the term *northerly turning error* is often used when describing the effects of centrifugal acceleration on compass magnet systems.

Turning through East or West

When turning from an easterly or westerly heading in either the northern or southern hemispheres (Figs 6.13 (e)–(h)) no error in indication results because the centrifugal acceleration acts in a vertical plane through the magnet system's point of suspension and centre of gravity. As will be noted the centre of gravity is merely deflected to the N or S of the point of suspension, thus increasing or decreasing the magnet system's pendulous resistance to dip.

A point which may be noted in connection with turns from E or W is that when the N or S end of the magnet system is tilted up, the line of the system is nearer to the direction where the directive force is zero, i.e. at right angles to the line of dip, and if the compass has not been accurately corrected during a compass "swing," any uncorrected deviating force will become dominant and so cause indications of apparent turns.

THE DIRECTIONAL GYROSCOPE

The directional gyroscope was the first gyroscopic instrument to be introduced as a heading indicator, and although for most modern aircraft it has been replaced by the synchronized remote-indicating compass, it is still used in a number of aircraft in its vacuum-driven form. The instrument employs a horizontal-axis gyroscope and, being non-magnetic, is used in conjunction with the magnetic compass; it defines the short-term heading changes during turns, while the magnetic compass provides a reliable long-term heading reference as in sustained straight and level flight. In addition, of course, the directional gyroscope overcomes the effects of magnetic dip, and of turning and acceleration errors inherent in the magnetic compass.

In its basic form the instrument consists of an outer ring pivoted about the vertical axis ZZ_1, and carrying a circular card graduated in degrees. The card is referenced against a lubber line fixed to the gyroscope frame. When the rotor is spinning, the gimbal system and card are stabilized so that by turning the frame the number of degrees through which it is turning may be read on the card against the lubber line.

The manner in which this simple principle is applied to practical instruments is governed by the manufacturer's design, but we may consider the version illustrated in Fig 6.14, and used in some light aircraft not equipped with remote-indicating compasses.

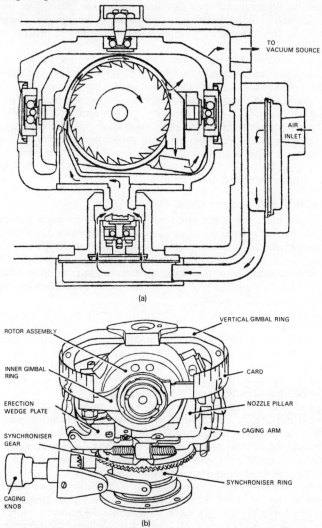

Fig 6.14 Air-driven directional gyroscope
(a) Airflow through instrument   (b) Gimbal and gyro rotor assembly

The rotor is enclosed in a case, or shroud, and supported in an inner ring which is free to turn about a horizontal axis YY_1. The inner ring is mounted in the vertical outer ring which carries the compass card and is pivoted on a vertical axis ZZ_1. The bearings for this ring are located in the top and bottom part of the instrument case, which thus forms the gyroscope frame.

The front of the case contains a cut-out through which the card is visible, and also the lubber line against which the card is referenced.

When the vacuum system is in operation a partial vacuum is created inside the case so that the surrounding air can enter through the filtered inlet and pass through channels in the gimbal rings to emerge finally through jets. The air jets impinge on the rotor "buckets," causing it to rotate at speeds between 12,000 and 18,000 rev/min.

To set the instrument so that it indicates the same heading as the magnetic compass, a caging and setting knob is provided at the front of the case. When this knob is pushed in, an arm is lifted thereby locking the inner ring at right angles to the outer ring, and at the same time meshing a bevel gear on the end of the caging-knob spindle with another bevel gear integral with the outer ring. Thus, a heading can be set by rotating the caging knob and the complete gimbal system. Once the correct setting is made the gyroscope is freed by pulling the caging knob out. The reason for caging the inner ring is to prevent it from precessing when the outer ring is rotated, and to ensure that, on uncaging, its axis is at right angles to the outer ring axis.

Control of Drift in Directional Gyroscopes

Drift is a fundamental characteristic of all directional gyroscopes and may be divided into two categories: (i) drift of the rotor axis from its horizontal position, caused by mechanical disturbances such as bearing friction and gimbal unbalance within the instrument, and (ii) drift of the gimbal system about its vertical axis due to the effect of the earth's rotation, referred to as *earth-rate error*. The effects of both forms of drift are controlled by erection devices and gimbal-ring balancing respectively.

Erection Devices

Erection devices form part of the rotor air-drive system and are so arranged that they sense misalignment of the rotor axis in terms of an unequal air reaction. In the instrument already described, after spinning the rotor the air exhausts through an outlet in the periphery of the rotor case and is directed on to a wedge-shaped plate secured to the outer ring.

In Fig 6.15 (a), the rotor axis is shown horizontal, and so the exhaust air outlet is upright and directly over the high point of the wedge located on the centre-line of the outer ring. The wedge therefore divides equally the air flowing from the outlet, and the air reaction applies horizontal forces R_1 and R_2 to the faces of the wedge. Since these forces are equal and opposite no torque is applied to the outer ring and the rotor axis remains horizontal.

When the rotor axis is tilted from the horizontal position the air outlet is no longer bisected by the wedge plate and a greater amount of air strikes one face of the wedge. From Fig 6.15 (b), it can be seen that the horizontal force R_1 is now greater than R_2 and a torque will therefore be applied in the direction of R_1 about the vertical axis ZZ_1. This torque may be visualized as being applied to the rim of the rotor at point F, causing the rotor to precess from this point, about axis YY_1 until its axis is horizontal and forces R_1 and R_2 again equal.

If the rotor tilts to such an angle that the exhaust air is entirely off the wedge plate, a secondary erection torque will be produced by the air stream issuing from the rotor spinning jet. In the normal horizontal attitude of the rotor the air stream impinges on the centre of the buckets as shown in Fig 6.15 (c). In the excessive tilt condition, however, the air stream strikes nearer to one side of the buckets and produces a force F which can be visualized as being a force F_1, acting at the point shown at (d). This force produces corresponding precession at point P_1 to return the rotor to its normal horizontal attitude. The force F will

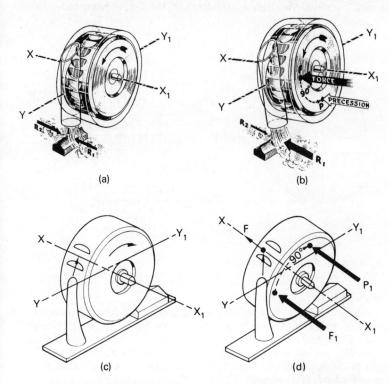

Fig 6.15 Knife-edge-type erection device
(a) Rotor axis horizontal   (c) Secondary erection torque
(b) Rotor axis tilted   (d) Precession from excessive tilt position

diminish as the rotor returns, but by this time the exhaust stream will again be in contact with the wedge plate so that final erection will be achieved in the manner indicated at (b).

Gimbal Ring Balancing

The drift due to the earth-rate error is independent of the other factors influencing drift, and its magnitude is governed by the relation

Drift = Earth's angular velocity × Sine of latitude

Calculation and plotting of the drift from the equator to the poles, and for both N and S latitudes, produces the graph shown in Fig 6.16.

The method of controlling this category of drift is to deliberately unbalance the inner ring so that a constant torque and precession are applied to the gimbal system. The unbalance is effected by a nut fastened to the rotor housing, and adjusted during initial calibration to apply sufficient torque and precession of the outer ring to cancel out the drift at the latitude in which it is calibrated. For all practical purposes this adjustment is quite effective up to 60° of latitude on the earth's surface. Above these latitudes the balancing nut has to be readjusted.

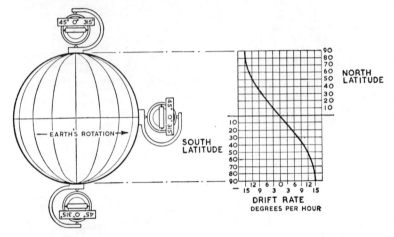

Fig 6.16   Drift due to earth's rotation

Gimballing Errors

Gimballing errors are those which can occur in the heading indications of all directional gyroscopes when an aircraft simply banks, turns or changes its attitude in pitch. *They are due solely to the gimbal system geometry when the gimbal rings are not mutually at right angles*, and are dependent upon: (i) the angles of climb, descent or bank, and (ii) the angle between the rotor axis and fore-and-aft axis of the aircraft.

In Fig 6.17, a series of diagrams illustrates the gimbal system geometry when an aircraft is in particular attitudes. At (*a*) the aircraft is represented as flying straight and level on an Easterly heading, and as the design of the gimbal system geometry is such that the rotor axis lies along the N-S axis, the three axes of the gimbal system are mutually at right angles, and the directional gyroscope will indicate the aircraft's heading without gimbal error. The same would also be true, of course, if the aircraft were flying on a Westerly heading.

If the aircraft banks to the left or right on either an Easterly or Westerly heading, or executes a left or right turn, the outer gimbal ring will be carried with the aircraft about the axis of the stabilized inner gimbal ring (diagram (*b*)). In this condition also the instrument would indicate, without gimbal error, the cardinal heading or change of heading during a turn.

At (*c*) the aircraft is assumed to be descending so that, in addition to the outer gimbal ring being tilted forward about the rotor axis, the inner gimbal also rotates, both rings maintaining the same relationship to each other. Again, there is no gimbal error; this would also apply in the case of an aircraft in a climbing attitude.

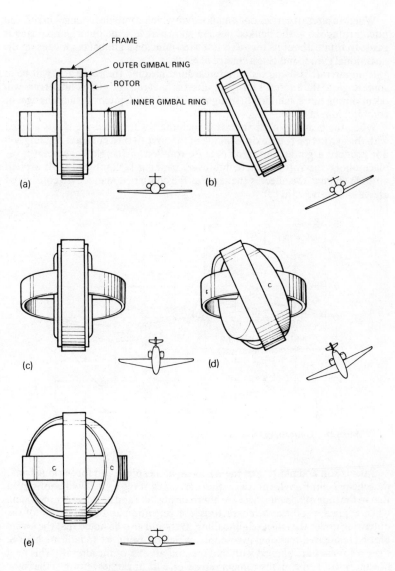

Fig 6.17 Gimballing errors

(a) Aircraft flying straight and level on Easterly heading: no error
(b) Aircraft banked to port on Easterly heading: no error
(c) Aircraft descending on Easterly heading: no error
(d) Aircraft banking to port and descending: error introduced
(e) Aircraft flying on intercardinal heading

When an aircraft carries out a manoeuvre which combines changes in roll and pitch attitude, e.g. the banked descent shown at (d), the outer gimbal ring is made to rotate about its own axis, thus introducing a gimbal error causing the directional gyro to indicate a change of heading.

If an aircraft is flying on an intercardinal heading the rotor axis will be at some angle to the aircraft's fore-and-aft axis, as at (e), and gimballing errors will occur during turns, banking in straight and level flight, pitch attitude changes or combinations of these.

When the heading of an aircraft is such that its fore-and-aft axis is aligned with the gyroscope rotor axis, banking of the aircraft on a constant heading will not produce a gimballing error because rotation of the gimbal system takes place about the rotor axis. If, however, banking is combined with a pitch attitude change, the effect is the same as the combined manoeuvre considered above and as shown at (d).

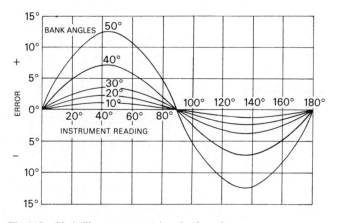

Fig 6.18 Gimballing errors at various bank angles

Calculation and plotting of the errors on all headings and for various bank, descent and climb angles produces the sine curves of Fig 6.18. It will be observed that in the four quadrants there are alternate positive and negative errors which, when applied, produce the characteristic acceleration and deceleration of the outer ring under the effect of gimballing. It should also be noted that the errors plotted relate to directional gyroscopes, which on being set to indicate N or S, their rotor axes are aligned with the fore-and-aft axis of the aircraft. This may not apply to all types of instrument; however, with any other setting of the rotor axis the error curves will merely be displaced relative to the aircraft heading scale, i.e. left or right by an amount equivalent to the angle between the rotor axis and the NS axis. At small angular displacements of the outer ring axis, the errors are small and diminish to zero when an aircraft returns to straight and level flight after executing a manoeuvre.

There is one final point which should be considered and that is the effect of the erection device whenever the angular relationship between the gimbal rings is disturbed. At all times this device will be attempting to re-erect the rotor into a new plane of rotation and will cause false erection, the magnitude of which depends on how long the erecting force is allowed to operate, i.e. the duration of the manoeuvre. The magnitude of the erecting force itself will depend on the angle of the rotor to the erection device. Thus, we see that on completion of a manoeuvre, it is possible to have an error in indication due to false erection, and that during a manoeuvre, an error can be caused which is a combination of both gimballing effect and false erection.

QUESTIONS

6.1 Define the following: (i) flux density, (ii) inverse square law, (iii) reluctance, (iv) magnetic moment, (v) period of a magnet.

6.2 The torque acting on a magnet at angle θ to a magnetic field is given by (a) mH sin θ, (b) mH, (c) FM cos θ. Which of these statements is correct?

6.3 What do you understand by the terms "hard-iron" and "soft-iron" magnetism?

6.4 Define the following: (i) magnetic meridian, (ii) magnetic variation, (iii) isogonal lines, (iv) agonal lines.

6.5 Draw diagrams to illustrate the relationship between the earth's magnetic components and magnetic dip at the equator and at the magnetic poles.

6.6 Define the unit in which the values of the magnetic components are given.

6.7 (a) What is an "aperiodic" compass?
(b) How are the effects you have described obtained in an aircraft compass?

6.8 Explain how the effects of temperature change in an aircraft compass are compensated.

6.9 (a) Describe the magnet system of a typical aircraft compass.
(b) How is the effect of dip overcome?

6.10 Describe a grid steering system and by means of diagrams explain the method of use.

6.11 (a) Define acceleration error and northerly turning error.
(b) Assuming an aircraft is flying in the southern hemisphere, what errors in compass readings will be introduced when (i) the aircraft accelerates on an Easterly heading, (ii) the aircraft turns from a Southerly heading towards East.

6.12 Explain briefly the principle of operation of a directional gyro. (SLAET)

6.13 Why is it necessary for the gyroscope assembly of a directional gyro to be caged when setting a heading?

6.14 Explain how the rotor and inner gimbal ring of a directional gyro are erected to the level position.

6.15 Define "earth rate error" and explain how its effects are controlled.

6.16 (a) How are gimballing errors caused?
(b) Under what flight conditions is the gimbal system unaffected?

7 Remote-indicating Compasses

In their basic form remote-indicating compasses currently in use are systems in which a magnetic detecting element monitors a gyroscopic indicating element. This virtual combination of the functions of both magnetic compass and directional gyroscope was a logical step in the development of instrumentation for heading indication and led to the wide-scale use of such systems as the *distant reading compass* and the *Magnesyn compass* in Allied military aircraft of World War II. Although successfully contributing to the navigation of such aircraft, these systems were not entirely free of certain of the errors associated with magnetic compasses and directional gyroscopes, and furthermore there were certain practical difficulties associated with the synchronizing methods adopted. In order, therefore, to reduce all possible sources of error and to provide subsequent designs of compass systems with self-synchronous properties, new techniques had to be adopted. The most notable of these were: the changeover from a permanent-magnet type of detector element to one utilizing electromagnetic induction as part of an alternating-current synchro system; the application of electronics; and the application of improved gyroscopic elements and precession control methods.

The manner in which the foregoing techniques are applied to systems currently in use depends on the particular manufacturer, and for the same reason the number of components comprising a system may vary. However, the fundamental operating principles of the main components shown in the block diagram of Fig 7.1 remain the same and are dealt with in this chapter.

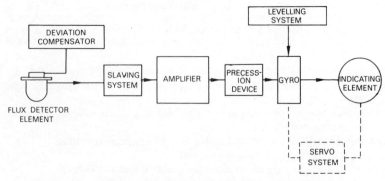

Fig 7.1   Essential components of a remote-indicating compass system

FLUX DETECTOR ELEMENTS

Unlike the detector element of the simple magnetic compass, the element used in all remote-indicating compasses is of the fixed type which detects the effect of the earth's magnetic field as an electromagnetically induced voltage and monitors a heading indicator by means of a variable secondary output voltage signal. In other words, the detector acts as an alternating-current type of synchro transmitter and is therefore another special application of the transformer principle (page 240).

In general, the construction of the element is as shown in Fig 7.2. It takes the form of a three-spoked metal wheel, slit through the rim between the spokes so that they and their section of rim act as three individual flux collectors.

Around the hub of the wheel is a coil corresponding to the primary winding of a transformer, while coils around the spokes correspond to secondary windings. The reason for adopting a triple spoke and coil arrangement will be made clear later in this chapter, but at this stage the operating principle can be understood by tracing through the development of only one of them.

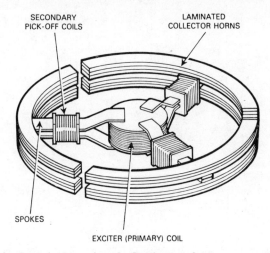

Fig 7.2   General construction of a flux detector element

We will first take the case of a single-turn coil placed in a magnetic field. The magnetic flux passing through the coil is a maximum when it is aligned with the direction of the field, zero when it lies at right angles to the field, and maximum but of opposite sense when the coil is turned 180° from its original position. Figure 7.3 (a) shows that for a coil placed at an angle θ to a field of strength H, the field can be resolved into two components, one along the coil equal to $H \cos \theta$ and the other at right angles to the coil equal to $H \sin \theta$. This latter component of the field produces no effective flux through the coil so that the total flux passing through it is proportional to the cosine of the angle between the coil axis and the direction of the field. In graphical form this may be represented as at (b).

If the coil were positioned in an aircraft so that it lay in the horizontal plane with its axis fixed on or parallel to the aircraft's fore-and-aft line, then it would be affected by the earth's horizontal component and the flux passing through the coil would be proportional to the magnetic heading of the aircraft. It is therefore apparent that in this arrangement we have the basis of a compass system able to detect the earth's magnetic field without the use of a permanent magnet.

180 *Aircraft Instruments*

Unfortunately, such a simple system cannot serve any practical purpose because, in order to determine the magnetic heading, it would be necessary to measure the magnetic flux and there is no simple and direct means of doing this. However, if a flux can be produced which changes with the earth's field component linked with the coil, then we can measure the voltage induced by the changing flux, and interpret the voltage changes so obtained in terms of heading changes. This is achieved by adopting the construction method shown in Fig 7.4.

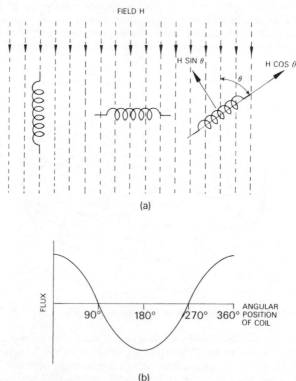

Fig 7.3 Single coil in a magnetic field
(*a*) Components   (*b*) Total flux

Each spoke consists of a top and bottom leg suitably insulated from each other and shaped so as to enclose the central hub core around which the primary coil is wound. The secondary coil is wound around both legs.

The material from which the legs of the spokes are made is Permalloy, an alloy especially chosen for its characteristic property of being easily magnetized but losing almost all of its magnetism once the external magnetizing force is removed. With this arrangement there are two sources of flux to be considered: (i) the alternating flux in the legs due to the current flowing in the primary coil; this flux is of the same frequency as the current and proportional to its amplitude; (ii) the static flux due to the earth's component H, the maximum value of

Remote-indicating Compasses 181

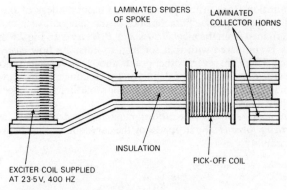

Fig 7.4 Vertical section of a detector-element spoke

which depends upon the magnitude of H and the cosine of the angle between H and the axis of the detector.

If we consider first that the axis of the detector lies at right angles to H, the static flux linked with the coils will of course be zero. Thus, with an alternating voltage applied to the primary coil, the total flux linked with the secondary will be the sum of only the alternating fluxes in the top and bottom legs and must therefore also be zero as shown graphically in Fig 7.5.

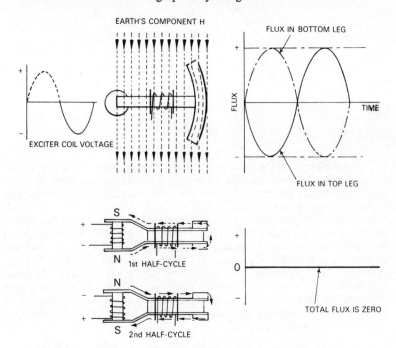

Fig 7.5 Total flux when detector spoke is at right angles to earth's field

The transition from primary coil flux to flux in the legs of the detector is governed by the magnetic characteristics of the material, such characteristics being determined from the *magnetization* or *B/H curve*. In Fig 7.6 the curve for Permalloy is compared with that for iron to illustrate how easily it may be magnetized. There are several other points about Fig 7.6 which should also be noted because they illustrate the definitions of certain terms used in connection with the magnetization of materials, and at the same time show other advantages of Permalloy. These are:

1. *Permeability*: which is the ratio of magnetic flux density B to field strength or *magnetizing force H*; the steepness of the curve shows that Permalloy has a high permeability.

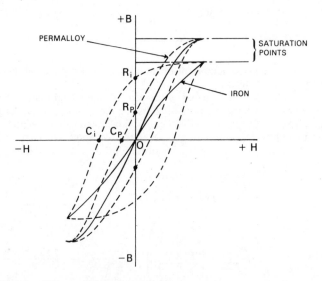

Fig 7.6 *B/H* curve and hysteresis loops

2. *Saturation point*: the point at which the magnetization curve starts levelling off, indicating that the material is completely magnetized. Note also that Permalloy is more susceptible to magnetic induction than iron as shown by its higher saturation point.

3. *Hysteresis* curve and loop*: these are plotted to indicate the lagging behind of the induced magnetism when, after reaching saturation, the magnetizing force is reduced to zero from both the positive and negative directions, and also to determine the ability of a material to retain magnetism. The magnetism remaining is known as *remanence* or remanent flux density and it will be noted that for iron this is very high (distance OR_i) thus making for good permanent magnetic properties. For detector elements, a material having the lowest possible remanence is required, and as the distance OR_P indicates, Permalloy meets this requirement admirably.

* The lagging of an effect behind its cause (from Greek *husteresis*, coming after).

4. *Coercivity*: this refers to the amount of negative magnetizing force (*coercive force*) necessary to completely demagnetize a material and is represented by the distances OC_1 and OC_P. Coercivity and not remanence determines the power of retaining magnetism.

In order to show the character of the flux waves produced in the legs of the detector, a graphical representation is adopted which is similar to that used for indicating electron tube characteristics. As may be seen from Fig 7.7 the waveshapes of the alternating primary fields are drawn across the flux density axis B

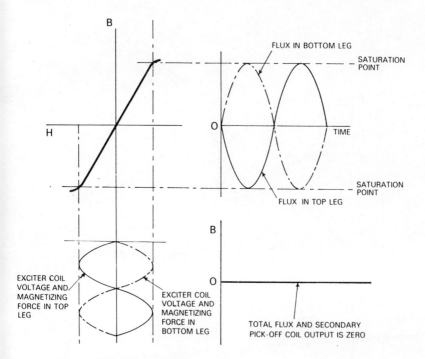

Fig 7.7 Flux wave characteristics

of the B/H curve, and those of the corresponding flux densities in the legs are then deduced from them by projection along the H axis. The total flux density produced in the legs is the sum of the individual curves, and with the detector at right angles to the earth's horizontal component H then, like the static flux linked with the secondary coil, it will be zero. Since the total flux density does not change, the output voltage in the secondary coil must also be zero.

Let us now consider the effects of saturation when the flux detector lies at any angle other than a right angle to the horizontal component H as indicated in Fig 7.8 (*a*). The alternating flux due to the primary coil changes the reluctance of the material thus allowing the static flux due to component H to flow into and out of the spoke in proportion to the reluctance changes. During those parts of the primary flux cycle when the reluctance is greatest, the static flux links with the secondary coil and the effect of this is to displace the axis, or datum, about which the magnetizing force alternates. The amount of this displacement depends upon the angle between the earth's field component H and the flux detector axis. This is shown graphically at (b).

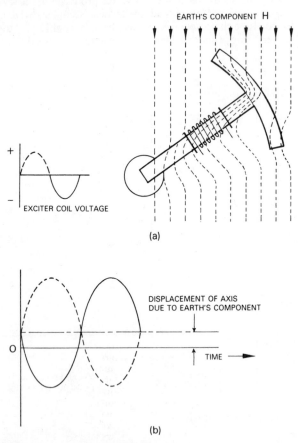

Fig 7.8 Effect of earth's component H
(a) Detector at an angle to component H
(b) Displacement of axis due to static flux

If we now apply a graphical representation similar to Fig 7.7 and include the static flux due to component H, the result will be as shown in Fig 7.9. It should be particularly noted that a flattening of the peaks of the flux waves in each leg of a spoke has been produced. The reason for this is that the amplitude of the primary coil excitation current is so adjusted that, whenever the datum for the magnetizing forces is displaced, the flux detector material is driven into saturation. Thus a positive shift of the datum drives the material into saturation in the direction shown, and produces a flattening of the positive peaks of the fluxes in a spoke. Similarly, the negative peaks will be flattened as a result of a negative shift driving the material into saturation at the other end of the B/H curve. The total flux linked with the secondary coil is as before, the sum of the fluxes in each leg, and is of the waveshape indicated also in Fig 7.9. When the flux detector is turned into other positions relative to the earth's field, then dependent on its heading the depressions of the total flux value become deeper and shallower.

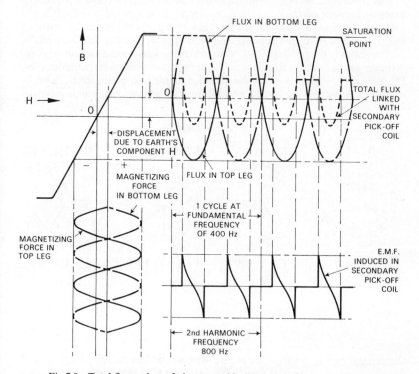

Fig 7.9   Total flux and e.m.f. due to earth's component H

Thus, the desired changes of flux are obtained and a voltage is induced in the secondary coil. The magnitude of this induced voltage depends upon the change of flux due to the static flux linked with the secondary coil which, in turn, depends upon the value of the effective static flux. As pointed out earlier, the value of the stasic flux for any position of the flux detector is a function of the cosine of the magnetic heading; thus the magnitude of the induced voltage must also be a measure of the heading.

One final point to be considered concerns the frequency of the output voltage and current from the secondary pick-off coil and its relationship to that in the primary excitation coil. During each half-cycle of the primary voltage, the reluctance of the flux detector material goes from minimum to maximum and back to minimum, and in flowing through the material the static flux cuts the pick-off coil twice. Therefore, in each half-cycle of primary voltage, two surges of current are induced in the pick-off coil, or for every complete cycle of the primary, two complete cycles are induced in the pick-off coil.

The supply for primary excitation has a frequency of 400 Hz; therefore the resultant e.m.f. induced in the secondary pick-off coil has a frequency of 800 Hz, as shown in Fig 7.9, and an amplitude directly proportional to the earth's magnetic component in line with the particular spoke of the detector element.

Having studied the operation of a single flux detector the reasons for having three may now be examined a little more closely. If we again refer to Fig 7.3, and also bear in mind the fact that the flux density is proportional to the cosine of the magnetic heading, it will be apparent that for one flux detector there will be two headings corresponding to zero flux and two corresponding to a maximum. Assuming for a moment that we were to connect an a.c. voltmeter to the detector, the same voltage reading would be obtained on the instrument for both maximum values because the instrument cannot take into account the direction of

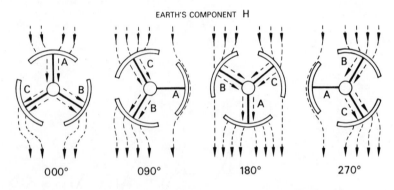

Fig 7.10 Path of earth's field through a detector and signals induced in pick-off coils

the voltage. For any other value of flux there will be four headings corresponding to a single reading of the voltmeter. However, by employing three simple flux detectors positioned at angles of 120° to one another, the paths taken by the earth's field through them, and for 360° rotation, will be as shown in Fig 7.10. Thus, varying magnitudes of flux and induced voltage can be obtained and related to all headings of the detector element without ambiguity of directional reference. The resultant voltage of the three detectors at any time can be represented by a single vector which is parallel to the earth's component H.

Figure 7.11 is a sectional view of a typical practical detector element. The spokes and coil assemblies are pendulously suspended from a universal joint which allows a limited amount of freedom in pitch and roll, to enable the element to sense the maximum effect of the earth's component H. It has no freedom in azimuth. The case in which the element is mounted is hermetically sealed and partially filled with fluid to damp out excessive oscillations of the element. The complete unit is secured to the aircraft structure (in a wing or vertical stabilizer tip) by means of a mounting flange which has three slots for mounting screws. One of the slots is calibrated a limited number of degrees on each side of a zero position corresponding to an aircraft installation datum for adjustment of coefficient A (see Chapter 8). Provision is made at the top cover of the casing for electrical connections and attachment of a deviation compensating device.

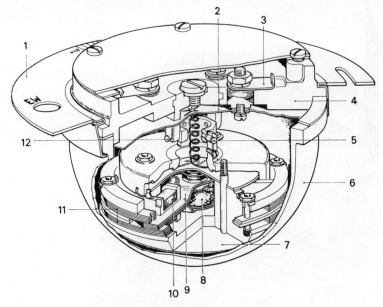

Fig 7.11 Typical flux detector element

1 Mounting flange (ring seal assembly)
2 Contact assembly
3 Terminal
4 Cover
5 Pivot
6 Bowl
7 Pendulous weight
8 Primary (excitation) coil
9 Spider leg
10 Secondary coil
11 Collector horns
12 Pivot

DIRECTIONAL GYROSCOPE ELEMENTS

Depending on the type of compass system and application of synchronized heading information, the directional gyroscope element may be contained within a panel-mounted indicator, or it may form an independent master unit located at a remote point and transmitting information to a slave indicator. Systems adopting the *master gyroscope reference* technique have many advantages; for example a number of synchronous heading transmission links can more conveniently be provided from a central source, multiple installations of panel-mounted instruments are eliminated, instruments may be mounted on sloping panels whereas integral gyroscope instruments always require vertical mounting.

Typical examples of both methods are shown in Figs 7.12 and 7.13. The gyroscope of the integral instrument is coupled, by means of gearing, to the rotor of a synchro which is supplied with signals from the flux detector element. The shaft on which the rotor is mounted also carries the heading card or in some cases a heading pointer. Thus, when relative movement between the gyroscope

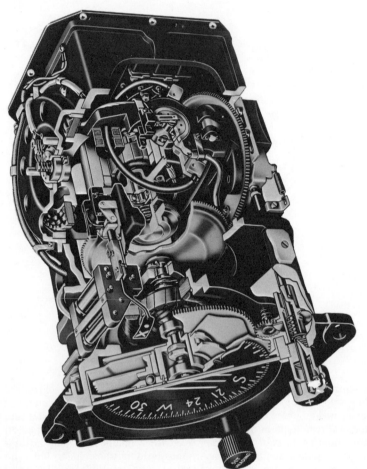

Fig 7.12 Typical integrated directional gyro element

and indicator case occurs, the card indicates a change of heading and the synchro rotor position within the stator is changed. Synchronizing of the gyroscope with flux-detector heading signals is achieved by the direct application of torques from a precession coil system controlled by amplifying the signals induced in the synchro rotor.

Fig 7.13 Typical master directional gyro element
    1 Housing for slave heading and data synchros
    2 Heading dial and vernier scale
    3 Levelling torque motor stator
    4 Rotorace bearing drive motor
    5 Levelling switch

The gyro rotor of the master unit is spherical in shape and supported in the inner gimbal ring which is fitted with hemispherical covers totally enclosing the gyro motor. This assembly is filled with helium and hermetically sealed. The rotors of the main slave heading synchro and an additional data synchro are mounted at the top of the outer gimbal ring. The stator of a levelling torque motor is attached to the bottom of the outer gimbal ring and its rotor is attached to the gyro casing. The motor is controlled by a liquid level switch secured to

190   Aircraft Instruments

the inner gimbal ring shaft. A heading dial is also fixed to the outer gimbal ring and is referenced against a fixed vernier scale. Both scales are viewed through an inspection window in the gyro outer casing, their purpose being to provide an arbitrary heading indication when testing the system in an aircraft. The complete assembly is mounted on antivibration mountings contained in a base which provides for attachment at the required location, and also for the connection of the necessary electrical circuits.

GYROSCOPE AND INDICATOR MONITORING

Monitoring refers to the process of reproducing the directional references established by the flux detector element as quantitative indications on the heading indicator. The principle of monitoring is basically the same for all types of compass systems and may be understood by reference to Fig 7.14.

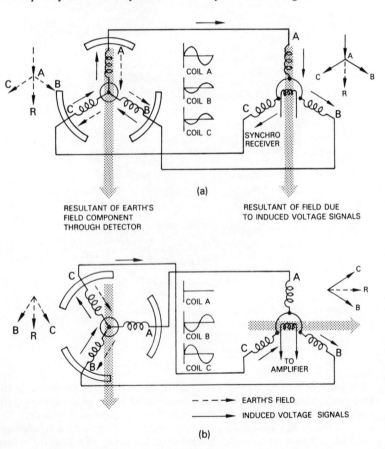

Fig 7.14   Directional gyro monitoring
(a) Heading = 000°   (b) Heading = 090°

When the flux detector is positioned steady on one heading, say 000°, then a maximum voltage signal will be induced in the pick-off coil A, while coils B and C will have signals of half the amplitude and opposing phase induced in them. These signals are fed to the corresponding tappings of the synchro receiver stator, and the fluxes produced combine to establish a resultant field across the centre of the stator. This resultant is in exact alignment with the resultant of the earth's field passing through the detector. If, as shown in Fig 7.14 (a), the synchro receiver rotor is at right angles to the resultant flux then no voltage can be induced in the windings. In this position, the synchro is at "null" and the directional gyro will also be aligned with the earth's resultant field vector and so the heading indicating element will indicate 000°. Now, let us suppose that we are able to turn the flux detector element independently of the gyroscope through 90° say; the disposition of the detector pick-off coils will be as shown at (b). No signal voltage will be induced in coil A, but coils B and C have increased voltage signals induced in them, the signal in C being opposite in phase to what it was in the "null" position. The resultant flux across the receiver synchro stator will have rotated through 90°, and because the gyroscope is still in its original position, the flux will now be in line with the synchro rotor and will therefore induce maximum voltage in the rotor. This error voltage signal is fed to a slaving amplifier in which its phase is detected, and after amplification is fed to the precession device, which precesses the gyroscope and synchro rotor in the appropriate direction until the rotor reaches the "null" position, once again at right angles to the resultant field. The heading indicating element will therefore be actuated to indicate the new direction taken up by the detector element, namely 090°. With the aid of Figs 7.10 and 7.14 the reader should have no difficulty in working out the directions of voltage signals and resultant fluxes for other positions of the detector element.

In the foregoing description, we have supposed relative movement between detector element and directional gyro merely as an aid to understanding how the gyroscope is monitored to the magnetic directional reference. What happens when the compass system is installed in an aircraft which turns through 90°? Fundamentally the operation is much the same; flux changes occur in the detector element, induced signal voltages flow through the receiver synchro stator and set up a rotating resultant flux. But what we must now realize is that in addition to the detector element, the gyro unit, as a complete assembly, is fixed to the aircraft. Thus, when the aircraft turns the gyro unit case will turn with it about the vertical axis of the gyroscope in exactly the same manner as the simple directional gyroscope. Through the medium of the drive mechanism between the gyroscope and synchro receiver rotor, the rotor will also turn relative to its stator, but as the detector element is turning, its signals are changing at the same rate as the synchro rotor rotates. Thus, during turns of an aircraft the synchro always remains effectively at the "null" position, no error voltage signals are produced, and so synchronism between gyroscope and detector element is maintained.

Figure 7.15 illustrates how the monitoring signal process is applied to a compass system employing an integral gyro indicator. In a system employing a master gyroscope the process is the same, but as the indicator is a separate unit a power follow-up synchro system (see page 244) is required to monitor both the gyroscope and the indicator.

The system is made up of servo synchros, a servo amplifier and a servo motor

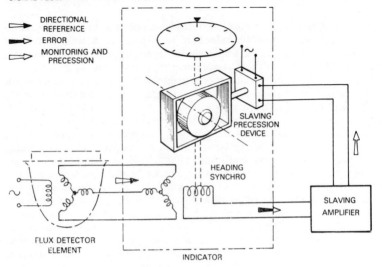

Fig 7.15 Integral gyro indicator monitoring

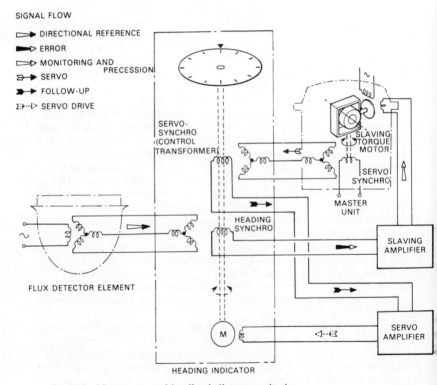

Fig 7.16 Master gyro and heading indicator monitoring

contained within the indicator, and a torque motor precession device (see page 195). The synchronous link between these components is shown in Fig 7.16, and from this it will also be noted that the servo motor is mechanically coupled to the two indicator synchros and the heading card.

Precession of the gyroscope to a directional reference drives the rotor of its servo synchro and produces a signal which is fed to the stator of the indicator servo synchro (a servo control transformer). The signal developed at the rotor flows to the servo amplifier and the output from this unit drives the servo motor in the appropriate direction. The rotors of the two indicator synchros are therefore driven to "null" positions, and the gyroscope and heading card are positioned at the directional reference established by the flux detector.

Methods of Precessing Gyroscopes to Directional References

Reference has already been made to the fact that synchronous monitoring of a gyroscope to established headings involves the application of the principles of gyroscopic precession. The construction and operation of the devices utilizing these principles vary, but the methods adopted in the gyroscopic components already described may be considered typical examples.

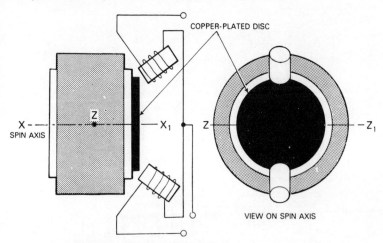

Fig 7.17 Eddy current method of precession

Figure 7.17 illustrates the device employed in an integral gyroscope instrument (see Fig 7.12). It consists of two coils mounted symmetrically on the outer gimbal ring and on either side of the horizontal spin axis of the gyroscope. The coils are supplied with signals from the slave heading synchro via the relevant amplifier output channel. The end surface of the gyro rotor housing adjacent to the coils is copper plated.

In the synchronized position of the gyroscope, the currents in the two coils are equal and they produce equal magnetic fields. As the copper-plated end surface of the rotor housing rotates past the coils, the magnetic fields induce eddy currents in the surface. The interaction between the fields and eddy currents produces equal torques on the rotor housing in opposite directions so that no movement of the outer gimbal takes place.

When a change of heading occurs, and as the gyroscope tends to drift out of synchronism, the current in one coil will be greater than that in the other, resulting in a greater interaction and torque on one or other side of the rotor. The axes of the coils lie in the plane through the axis on which the rotor housing is pivoted to the outer gimbal ring. The greater torque due to the greater current will therefore act on the appropriate side of the ring, causing it to precess directly about its vertical axis. As it precesses it turns the rotor of its synchro until the "null" position is reached and the torques produced by the two currents are once again in balance.

A second method employed in another type of integral gyro instrument is shown in Fig 7.18. It is operated by alternating current and consists of a copper-plated Mumetal ring attached to the gyro rotor housing and passing through a double coil assembly on the outer gimbal ring. One coil is energized with a reference voltage of approximately 40 V from a transformer, while the other coil

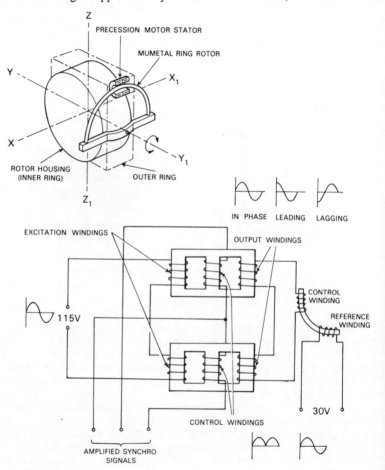

Fig 7.18 A.C. precession motor system

is supplied with a controlling voltage from a pair of magnetic amplifiers, the d.c. control windings of which are connected to the signal amplifier. This arrangement may thus be considered as a special type of two-phase torque motor.

At the "null" position of the synchro and with the gyroscope synchronized with the magnetic heading, no error signal is fed to the amplifier and the control windings of the magnetic amplifiers carry equal currents. Under these conditions the amplifiers can be considered as two identical transformers connected in opposition so that no voltage is applied to the control winding of the precession coil.

When the gyroscope is out of synchronism, the error signal fed to the signal amplifier results in a differential output to the d.c. control windings of the magnetic amplifiers, the output to one increasing while the output to the other is decreasing. Now, when direct current is applied to the control winding of a magnetic amplifier, the magnetic flux induced in the core laminations alters the transformer effect between the excitation and output winding in such a manner that the output is reduced for an increased control-winding current, and increased for a reduced control-winding current. Therefore, when the gyroscope and its synchro rotor are out of synchronism, the output of the magnetic amplifier receiving the lower d.c. control signal will supply a greater output voltage to the precession-coil control winding, causing current to flow through it.

The output is related in phase to the directional displacement of the gyroscope and synchro rotor from their synchronized positions; thus the precession-coil control winding voltage will either lead or lag the reference winding voltage. The field due to the coil current interacts with eddy currents induced in the Mumetal ring, and there is a tendency for it to move. It is, however, attached to the gyro rotor housing; therefore a reactive torque is set up which precesses the gyroscope and synchro rotor back to the synchronized heading and restores equal values of direct current in the magnetic-amplifier control windings. The maximum precession rate is between $2°$ and $5°$ per minute.

A typical precession device used in the master gyro system is one employing a torque motor connected as shown in Fig 7.18. The motor is of the two-phase induction type, the stator reference phase being supplied with a fixed voltage signal while the control phase is supplied with the variable monitoring signals from the slaving synchro and amplifier. The stator is fixed to the gyro outer gimbal ring and the rotor is fixed to the inner gimbal ring shaft.

When the gyroscope is out of synchronism the voltage signal from the amplifier passes through the control winding of the motor. Depending on the direction of gyroscope displacement the signal phasing will either lead or lag behind that of the reference signal. A rotating field is thus set up in the rotor of the torque motor which tends to rotate the gyroscope rotor housing about its pivoted axis. This is resisted by the gyroscope so that the torque created is exerted on the outer gimbal ring, causing it to precess directly about the vertical axis until the "null" position of the synchro system is again established.

GYROSCOPE LEVELLING

In addition to the use of efficient synchronous transmission systems, it is also essential to employ a system which will maintain the spin axis of the gyroscope in a horizontal position at all times. This is a requirement for all directional

196 *Aircraft Instruments*

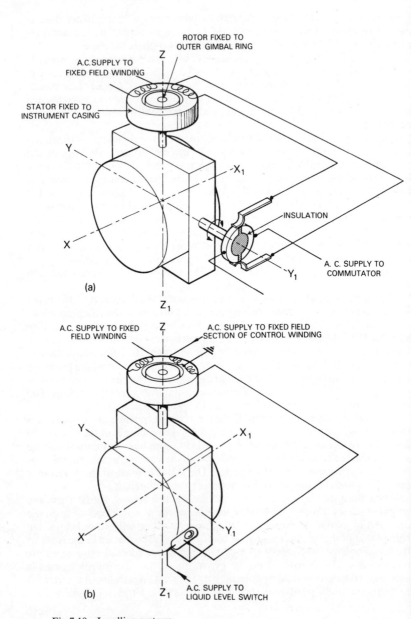

Fig 7.19 Levelling systems
(*a*) Commutator type   (*b*) Liquid-level switch type

gyroscopes (see also page 172), to overcome random drift of the gyro rotor housing due to bearing friction and the curved path followed by an aircraft over the earth's surface.

The levelling or erection systems utilized in the gyroscopes of remote-indicating compasses are similar in design and operating principle and consist of a switch and a torque motor. The switch may be either of the double-segment commutator type or the liquid-level type mounted on the gyro rotor housing so as to move with it. The torque motor is a two-phase induction motor usually located so that its stator is attached to the bottom of the outer gimbal ring, and its rotor is attached to the gyro casing.

The electrical interconnection of levelling system components is shown in Fig 7.19. The commutator switch shown at (*a*) is made up of two segments suitably insulated from each other and continuously supplied with a low alternating voltage. Two brushes bear against the commutator and are fixed to the outer gimbal ring and electrically connected to each end of the torque-motor control winding. In the level position of the gyro rotor housing, the brushes are in contact with the insulated portions of the commutator and no current flows through the control winding.

When the gyro rotor housing tilts in either direction, the "live" segments of the commutator come under the brushes and energize the control winding, thus inducing a field whose direction is dependent on the direction of tilt. A rotating field is produced by the current in the torque-motor stator winding, and the rotor in trying to follow it furnishes a torque on the outer ring to precess the gyro rotor housing back to the horizontal position.

The liquid-level switch shown in Fig 7.19 (*b*) consists of a small sealed glass tube containing a globule of mercury and two electrodes. The switch is secured to the gyro rotor housing and has one of its electrodes connected to a low alternating voltage source and the other connected to one end of the torque motor control winding. It will be noted from the diagram that the control winding, unlike that of the commutator switch system, is formed in two parts; one is continuously energized, and the other is energized only through the liquid-level switch. The fields set up by the currents in the two windings produce torques in the manner already described, but since the fields are in opposition to each other precession to the level position is due to a resultant torque.

Figure 7.20 illustrates a method used in a current type of master gyro unit for counteracting drift of the gimbal system due to bearing-friction torques. The inner gimbal ring is supported within the outer gimbal ring by two special bearings, known as Rotorace bearings. Each bearing consists of an inner race, a middle race, an outer race, and two sets of ball bearings; the middle race is constantly rotated by a drive motor working through a gear train.

Rotation of the middle race uniformly distributes around the axis of rotation any frictional torques present in the bearing. The gearing to the middle races of the two bearings is such that the races are rotated in opposite directions; thus the frictional torques of the two bearings oppose each other. Drift that may be introduced by mismatching of the bearings is cancelled by periodically changing the direction of rotation of the middle races. This is achieved by reversing the rotation of the drive motor through a switch which is operated by a cam. The cam, in turn, is driven by a worm gear from the drive shaft between the two bearings. Capacitors connected across the switch suppress any tendency for contact arcing to take place.

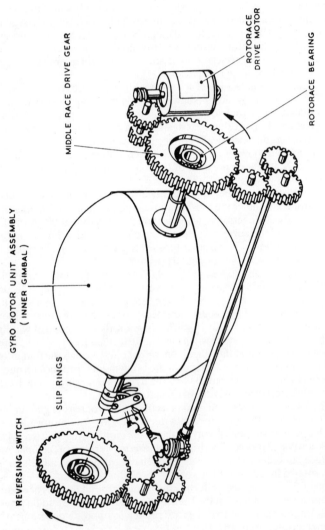

Fig 7.20 Rotorace bearing system

MODES OF OPERATING COMPASS SYSTEMS

All compass systems provide for the selection of two categories or modes of operation: *slaved*, in which the gyroscope is monitored by the flux detector element, and *free gyro*, or *D.G.*, in which the gyro is isolated from the detector to function as a straightforward directional gyroscope. The latter operating mode is selected whenever malfunctioning of directional reference-signal circuits occurs or when an aircraft is being flown in latitudes at which the horizontal component of the earth's magnetic field is an unreliable reference.

Synchronizing Indicators

The function of synchronizing indicators, or annunciators as they are sometimes called, is to indicate whether or not the gyroscope of a compass system is synchronized with the directional reference sensed by the flux detector element. The indicators may form an integral part of the main heading indicator or they may be a separate unit mounted on the main instrument panel. They are actuated by the monitoring signals from the flux detector and are connected into the gyro precession circuit.

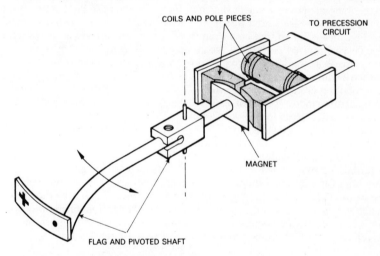

Fig 7.21 Typical synchronizing indicator

A typical integral annunciator is shown in Fig 7.21 and serves to illustrate the function of indicators generally. It consists of a small flag, marked with a dot and a cross, which is visible through a small window in one corner of the heading indicator bezel. The flag is carried at one end of a pivoted shaft. A small permanent magnet is mounted at the other end of the shaft and positioned adjacent to two soft-iron cored coils connected in series with the precession circuit.

When the gyroscope is out of synchronism, the monitoring and precession signals flow through the annunciator coils and induce a magnetic field in them. The field reacts with the permanent magnet causing it to swing the shaft to one side so that either the dot or the cross on the flag will appear in the annunciator window. The particular indication shown depends upon the direction in which the gyroscope is precessing and will remain until synchronism is regained.

Under synchronized conditions the annunciator window should ideally be clear of any image. During flight, however, the flag oscillates slowly between a dot and a cross indication. This is due to pendulosity effects on the flux detector element and does in fact serve as a useful indication that monitoring is taking place.

Manual Synchronization

When electrical power is initially applied to a compass system operating in the "slaved" mode, the gyroscope may be out of alignment by a large amount. It will, of course, start to synchronize, but as the gyroscopes normally have low rates of precession (1° to 2° per minute), some considerable time may elapse before synchronization is effected. In order to speed up the process, systems always incorporate a fast synchronizing facility which can be manually selected.

Various methods are used but all are based on the principle of driving a synchro to its "null" position. An example of one of the methods is that applied to the integral gyro instrument shown in Fig 7.12. The push knob at the front of the instrument is marked with a dot and a cross, and is coupled to a caging mechanism similar to that employed in an air-driven directional gyroscope. The desynchronized position of the gyroscope will be indicated by a dot or a cross image in the annunciator window, and to synchronize it the knob is pushed in and rotated in the appropriate direction. The gyroscope is caged, and as it rotates it drives the slaving synchro rotor to its "null" position. When the annunciator window is clear of an image the gyroscope is synchronized and the knob is released under the action of a spring.

The principle of a method applied to a system utilizing a master gyro may be understood from Fig. 7.13. The indicator also employs a synchronizing knob, but in this case it is coupled to the stator of the servo synchro. When the knob is pushed in and rotated in the direction indicated by the annunciator, the rotation of the stator of the servo synchro induces an error voltage signal in the rotor. This is fed to the servo amplifier and motor which drives the slave heading synchro rotor and gyroscope, via the slaving amplifier, into synchronism with the flux detector. At the same time, the servo synchro rotor is driven to the "null" position and all error signals are removed.

INDICATION OF MAGNETIC HEADING IN POLAR REGIONS

With the introduction of regular flight operations in the vicinity of and over the polar regions by military aircraft and many civil aircraft operators, the necessity for developing new navigational methods and equipment became of paramount importance. The main difficulties of navigation in these regions are associated with the following problems: the convergence of the meridians and isogonals, the long twilight periods limiting the use of astro navigation, magnetic storms and radio blackouts limiting the use of radio navigational aids, and limitations in the directional references established by magnetic compasses and some remote-indicating compass systems.

It will be recalled from Chapter 6, that the magnetic North cannot be defined as an exact point, but may be described as the area within which the horizontal component H of the earth's magnetic field is zero, the entire field being vertical. It is therefore clear that the use of a direct-reading compass or remote-indicating compass in this area cannot provide the required accurate directional reference. Another circumstance which makes magnetic steering impracticable in the polar regions is the fact that the deflection of the magnet system from magnetic North is very large and varies rapidly with distance.

The gyroscopes of present-day compass systems are designed to be substantially free from drift due to gimbal-ring bearing friction and unbalance, and so their use in the "free gyro" mode of operation offers an obvious solution to the

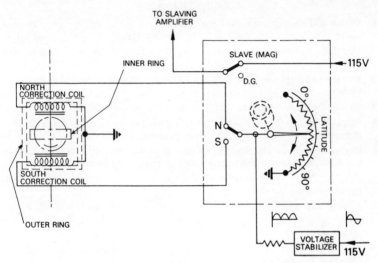

Fig 7.22   Latitude controlling circuit

problem. However, a gyroscope still has an earth-rate error which varies with the latitude in which it operates and is greatest in polar regions (see Fig 6.16). To obtain accurate heading information, therefore, the errors must be eliminated either by calculating the corrections required and applying them directly to indicated headings, or by incorporating an automatic correcting device in the gyroscope. In compass systems designed for polar flying the latter method is used and takes the form of a controlled precession circuit.

A typical circuit consists essentially of three main parts: (i) a switch for selecting the hemisphere in which a flight is being made, i.e. North or South, (ii) a selector knob, potentiometer and dial calibrated in degrees of latitude, thus providing for the setting of the latitude in which the aircraft is being flown, and (iii) two small d.c. electromagnets built into the outer ring, one on either side of the inner gimbal ring. The controlling circuit is illustrated in Fig 7.22.

When the hemisphere selector switch is set to the appropriate position it allows a high direct voltage from a special power unit to flow to the coil of the electromagnet selected. The current also passes through the potentiometer of the latitude compensator so that it can be adjusted to give the control required at the latitude selected. As the gyroscope drifts, the rotor turns through the magnetic field of the energized coil, and eddy currents are induced in the rotor with the result that a torque is applied to the inner gimbal ring, which precesses the complete gimbal system about its vertical axis at the same rate, but in opposition to its drift. The precession is counter-clockwise when the northern-hemisphere magnet is energized, and clockwise when the southern hemisphere magnet is energized. A voltage-stabilizing circuit ensures that the current through an energized coil is unaffected by variations in the direct voltage from the power supply unit.

QUESTIONS

7.1 Sketch and describe in detail the construction of a heading detector unit suitable for use in a remote-indicating compass system. (SLAET)

7.2 Sketch the hysteresis loops of soft-iron and Mumetal specimens and explain the shape of the loops. (SLAET)

7.3 By means of diagrams describe how fluxes and voltages are induced in a detector element.

7.4 Explain how unambiguous directional references are obtained.

7.5 What effect does the alternating current supplied to the exciter coil of a detector element have on the earth's field passing through the element?

7.6 Explain the effects you consider the pendulous mounting of a detector element will have on indicated headings during turns and accelerations.

7.7 Describe the general construction of a remote-indicating compass directional-gyro element.

7.8 Explain how a directional gyro element is monitored by the flux detector.

7.9 With the aid of a diagram describe the construction of a typical precession device and explain its operation.

7.10 (a) On which components of a directional gyro element are the rotor and stator of a levelling torque-motor fitted?
(b) Explain the operation of a typical levelling system.

7.11 (a) Under what conditions is the "D.G." mode of operation selected?
(b) Describe how indications of synchronism between the directional gyro and flux detector elements are obtained.

7.12 Describe how the "earth rate error" is automatically corrected in a polar compass system.

8   Aircraft Magnetism and its Effects on Compasses

A fact which has always been a challenge to the designers of aircraft compasses is that all aircraft are themselves in possession of magnetism in varying amounts. Such magnetism is, of course, a potential source of error in the indications of compasses installed in any type of aircraft and is unavoidable. However, it can be analysed and, for any aircraft, can be divided into two main types and also resolved into components acting in definite directions, so that steps can be taken to minimize the errors, or deviations as they are properly called, resulting from such components.

The two types of aircraft magnetism can be divided in the same way that magnetic materials are classified according to their ability to be magnetized, namely hard iron and soft iron (see also page 154).

Hard-iron magnetism is of a permanent nature and is caused, for example, by the presence of iron or steel parts used in the aircraft's structure, in power plants and other equipment, the earth's field "building" itself into ferrous parts of the structure during construction and whilst lying for long periods on one heading.

Soft-iron magnetism is of a temporary nature and is caused by metallic parts of the aircraft which are magnetically soft becoming magnetized due to induction by the earth's magnetic field. The effect of this type of magnetism is dependent on the heading and attitude of the aircraft and its geographical position.

There is also a third type of magnetism due to the sub-permanent magnetism of the "intermediate" iron, which can be retained for varying periods. Such magnetism depends, not only on the heading, attitude and geographical position of an aircraft, but also on the nature of its previous motion, vibrations, lightning strikes and other external agencies.

The various components which cause deviation are indicated by letters, those for permanent, hard-iron, magnetism being capitals, and those for induced, soft-iron, magnetism being small letters. It is also important to note at this juncture that positive deviations are termed Easterly and negative deviations Westerly.

EFFECTS OF MAGNETIC COMPONENTS ON COMPASSES

Components of Hard-iron Magnetism

The total effect of this type of magnetism at the compass position may be considered as having originated from bar magnets lying fore and aft, athwartships and vertically. These are shown schematically in Fig 8.1. In order to analyse their effects they are respectively denoted as *Components P, Q* and *R*. The strength of these components will not vary with heading or change of latitude but may vary with time due to a weakening of the magnetism in the aircraft. Referring again to Fig 8.1, we should note particularly when the Blue poles of the magnets are forward, to starboard, and beneath the compass respectively, the components are positive, and when the poles are in the opposite directions the components are negative. The fields always pass from the Blue pole to the Red pole.

When an aircraft is heading North, the equivalent magnet due to component P will, together with the compass, be in alignment with the aircraft's fore-and-aft axis and the earth's component H, and so will cause no deviation in compass heading. If we now assume that the aircraft is to be turned through 360°, then as it commences the turn, the compass magnet system will (ignoring pivot

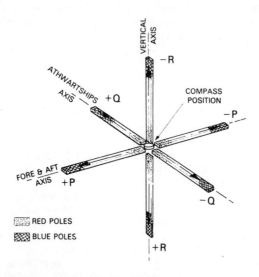

Fig 8.1   Components of hard-iron magnetism

friction, liquid swirl, etc.) remain attracted to the earth's component H. Component P, however, will align itself in the resultant position and will cause a deviation in the compass reading of magnetic North, making it read so many degrees East or West of North, depending on the polarity of the component. This deviation will increase during the turn and will be a maximum at East, then it will decrease to zero at South, increasing once again to a maximum but in the opposite direction at West, and then decreasing to zero when turned on to North once again. A plot of the deviations caused by a positive and a negative component P on all headings results in the sine curves shown in Fig 8.2.

Component Q produces similar effects, but since it acts in the athwartships axis, deviation is a maximum on North and South headings and zero on East and West headings. Plotting these deviations for positive and negative components, results in cosine curves shown in Fig 8.3; thus deviations due to Q vary as the cosine of the aircraft's heading.

Component R acts in the vertical direction, and when the aircraft is in its normal level flight attitude, it has no effect on the compass magnet system. If, however, the aircraft flies with either its fore-and-aft axis or athwartships axis in positions other than horizontal, then component R will be out of the vertical and will have a horizontal component affecting the compass magnet system.

Figure 8.4 (*a*) illustrates the horizontal components of component *R* when an aircraft is in the nose-up and nose-down attitudes. When the aircraft is in the nose-up attitude, the horizontal component of a positive component *R* acts in a forward direction, i.e. in the same sense as a positive component *P*, and when

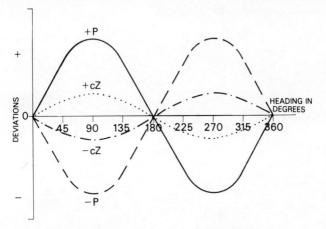

Fig 8.2   Deviation curves due to components *P* and *cZ*

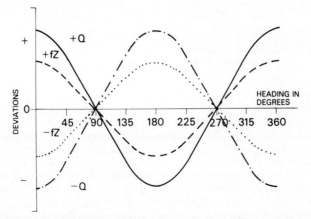

Fig 8.3   Deviation curves due to components *Q* and *fZ*

the aircraft is in the nose-down attitude the horizontal component acts rearwards in the same sense as a negative component *P*. Thus, in either of these attitudes *R* will cause maximum deviation on East and West headings.

The actual amount of deviation depends on the value of *R* and the angle between the aircraft's fore-and-aft axis and the horizontal. The resultant curves of deviations will be sine curves similar to those produced by component *P*.

If component *R* is negative, as may also be seen from Fig 8.4 (*a*), the horizontal components will act in the opposite directions.

When an aircraft is in a banked attitude as shown at (*b*), component *R* will again have a horizontal component with an effect similar to the athwartships component *Q*, causing maximum deviations on North and South. A positive component *R* produces an effect similar to a negative component *Q* when the aircraft is banked starboard wing down. When banked port wing down the effect is similar to a positive component *Q*.

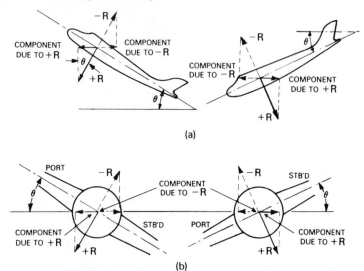

Fig 8.4 Component *R* and effects
(*a*) Effects in nose-up and nose-down attitudes
(*b*) Effects in banked attitude

If component *R* is negative the horizontal components will have the opposite effects. As before, the actual amounts of deviation depend on the magnitude of *R* and the angle between the aircraft's axis and the horizontal.

A question often raised in connection with component *R* is "Are the effects really serious enough to bother about?" As stated earlier, they depend on the magnitude of the component and the angles through which the aircraft's pitch and bank attitude is changed. Normally, the angle of climb and descent of most aircraft is reasonably shallow, and so deviations are usually small. In some aircraft with a large component *R* a large effect on deviation is experienced in the nose-up attitude. During turns an aircraft is banked, but here again angles and deviations are usually small. If direct-reading compasses are used in an aircraft they suffer far more from errors due to turns so that any additional errors resulting from component *R* effects are of little or no practical interest. The arrangement of detector elements of remote-indicating compasses is such that turn errors are eliminated and component *R* is negligible.

Components of Soft-iron Magnetism

The soft-iron magnetism which is effective at the compass position may be considered as originating from a piece of soft iron in which magnetism has been induced by the earth's magnetic field. This field has two main components *H*

Table 8.1

Earth's magnetic components	Aircraft axes to which related	Polarity
$H\begin{cases}X\\Y\end{cases}$	Fore and aft Athwartships	+ Forward—Aft + Starboard—Port
Z	Vertical	+ Down—Up

and Z, but in order to analyse soft-iron magnetism, it is necessary to resolve H into two additional horizontal components X and Y and to relate them, and the vertical component Z, to the three principal axes of an aircraft. This relationship and the polarities are given in Table 8.1.

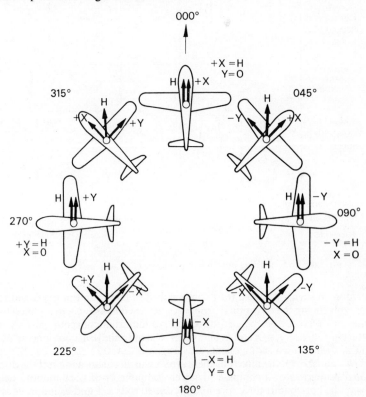

Fig 8.5 Changes of X, Y and Z components with aircraft heading changes

The polarities and strengths of components X and Y change with changes in aircraft heading because the aircraft turns relative to the fixed direction of component H. The changes occurring through the cardinal and quadrantal points are indicated in Fig 8.5.

Component Z acts vertically through the compass and therefore does not affect the directional properties of the compass magnet system.

If an aircraft is moved to a different geographical position, then because of the change in earth's field strength and direction, all three components will change. A change in the sign of component Z will only occur with a change in magnetic hemisphere when the vertical direction of the earth's field is reversed.

Each of the three earth's field components produces three soft-iron components which are visualized as resulting from induction in a number of soft-iron rods disposed along the three axes of the aircraft, and around the compass

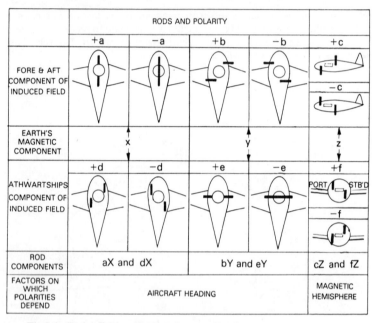

Fig 8.6 Rods affecting directional properties of a compass

position, in such a way that their combined effect is the same as the actual soft iron influencing the directional properties of the compass. There are therefore nine rods of soft-iron magnetism which are indicated conventionally by the letters a, b, c, d, e, f, g, h and k, and when related to the field components X, Y and Z, their soft-iron components are designated aX, bY, cZ; dX, eY, fZ; and gX, hY and kZ. Of the nine rods there are six which will always affect the directional properties of a compass since the components of their induced fields always lie horizontal: three fore-and-aft due to rods a, b and c, and three athwartships due to rods d, e and f. These are shown in pictorial and tabular form in Fig 8.6. In passing we may note that slight deviations could be produced by the vertically induced fields of rods g, h and k in a manner similar to those resulting from component R of hard-iron magnetism.

There are two main points which should be noted with reference to Fig 8.6. Firstly, each rod has alternative positions, designated positive and negative.

The reason for this is because each component of the aircraft's magnetism may act in one of two reciprocal directions. The polarity designation is dependent only on the position of the rods relative to the compass, and is not affected by any polarity which the rods may acquire as a result of being magnetized by the earth's field.

The second point to note is that with the exception of the $-a$ and $-e$ rods, there are two positions relative to the compass in which each rod is positive and two in which each is negative; thus the fields pass from the poles nearer the compass. We often refer to these fields as being due to asymmetric horizontal soft-iron.

The deviation curves due to rod components aX, eY, bY and dX are illustrated in Fig 8.7. Rod components cZ and fZ have the same effects as hard-iron components P and Q respectively. A further point to note in connection with cZ and fZ is that their polarities and direction depend on whether the aircraft is in the Northern or Southern hemisphere.

Total Magnetic Effects

The total magnetic effects of aX and eY are quadrantal and may be found by summating the individual curves of deviation shown in Fig 8.7. The combining of the curves of deviation due to components bY and dX produces the total effects shown in Fig 8.8, and from this it will be apparent that, in addition to producing deviation which varies as the cosine of twice the heading, components bY and dX may also produce constant deviation or a combination of both.

The total magnetic effect at a compass position is the sum of the earth's field components X, Y and Z, hard-iron components P, Q and R, and the respective soft-iron components acting along the three axes of the aircraft. The effects are shown schematically in Fig 8.9, together with the expressions for calculation of the combined forces represented as X', Y' and Z'. The signs used in the expressions indicate that the sums of the quantities are algebraic.

In the foregoing explanation of aircraft magnetism, its components and effects, no direct reference has been made to the detector elements of remote-indicating compasses. It should not be construed from this that such elements are entirely immune from extraneous magnetic fields; All compasses, regardless of their method of detecting the earth's component H, must suffer from the ultimate effect which is deviation. The only difference is the manner in which the aircraft's magnetic components affect the detecting elements and cause deviation.

The detector element of a direct-reading compass, as we already know, is of the freely suspended permanent-magnet type, and so deviation is caused by direct deflection of the element from magnetic North. This, of course, is not true of a remote-indicating compass detector element, so that when an aircraft turns the detector element turns with it and its associated hard-iron and soft-iron magnetic components. How then can deviation be produced?

The answer lies in the fact that the material of the detector element has a high permeability and is therefore very receptive to magnetic flux. It will be recalled that accurate heading indication is dependent on the displacement of the H axis, about which the magnetizing force alternates, by the earth's field component. Since the material is easily magnetized by this component, then it is just as easy for it to be magnetized by components of aircraft magnetism. Consequently,

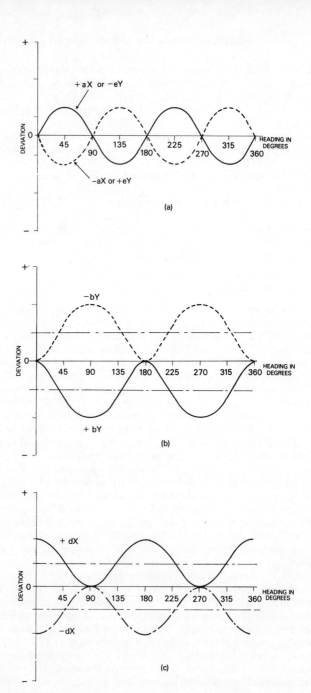

Fig 8.7 Deviation curves due to rod components
(a) Components aX and eY (b) Component bY (c) Component dX

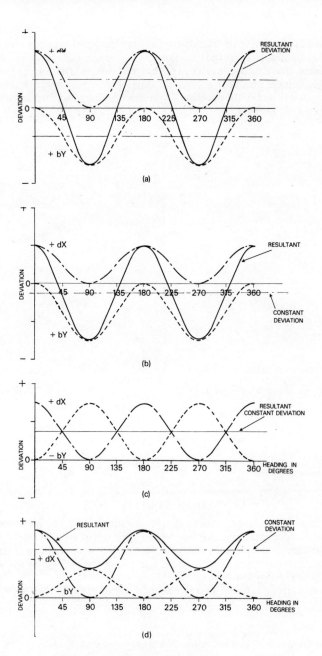

Fig 8.8 Total effects of components bY and dX
  (a) Components of same sign and equal value
  (b) Components of same sign and unequal value
  (c) Components of opposite sign and equal value
  (d) Components of opposite sign and unequal value

the *H* axis is displaced to a false datum in either the positive or negative direction. This, in turn, changes the total flux linking the secondary pick-off coils of the detector, thus causing an error in the induced voltage. The current resulting from this voltage produces a corresponding error signal in the heading or slaving synchro, and so the directional gyroscope and heading card are deviated from the correct heading.

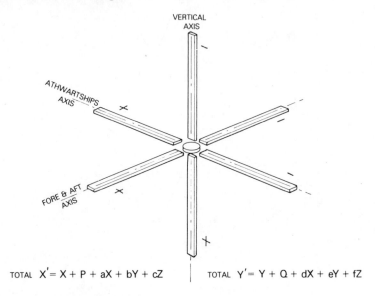

Fig 8.9 Total magnetic effect at compass position

DEVIATION COEFFICIENTS

Before steps can be taken to minimize the deviations caused by hard-iron and soft-iron components of aircraft magnetism, their values on each heading must be obtained and quantitatively analysed into *coefficients of deviation*. There are five coefficients named *A*, *B*, *C*, *D* and *E*, termed positive or negative as the case may be, and expressed in degrees.

The relationship between aircraft magnetism and the coefficients is shown in Fig 8.10.

Coefficient A
This represents a constant deviation due to the combined effects of components *bY* and *dX* of unlike signs.

Referring again to Fig 8.7 we note that components $-bY$ and $+dX$ both cause Easterly deviation, and a combination of these two will cause a coefficient $+A$; while components $+bY$ and $-dX$ both cause Westerly deviation and in

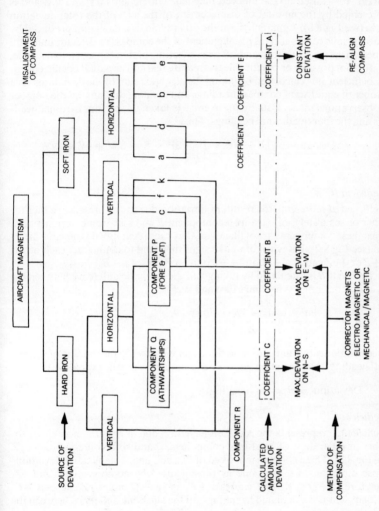

Fig 8.10 Relationship between aircraft magnetism and deviation coefficients

combination causes a coefficient $-A$. In each case, the two components, as regards their maximum effects on the compass, must be equal. Deviation coefficient A may be termed either *real* or *apparent*.

Real A is caused by the induced magnetic components bY and dX, and is represented by the amount of displacement of the axis of the total deviation curve (see Fig 8.8). Apparent A, on the other hand, is a deviation produced by non-magnetic causes such as misalignment of the compass or detector unit with respect to the aircraft's fore-and-aft axis, offsetting of the magnet system needle, etc. It is not possible or practicable to separate real and apparent coefficients A.

Coefficient A is calculated by taking the algebraic sum of the deviations on a number of equidistant compass headings and dividing this sum by the number of observations made. Usually the average is taken on the four cardinal headings and the four quadrantal headings. Thus

$$A = \frac{\text{Deviation on N + NE + E + SE + S + SW + W + NW}}{8}$$

Coefficient B

Coefficient B represents the resultant deviation due to the presence, either together or separately, of hard-iron component P and soft-iron component cZ. When these components are of like signs, they cause deviation in the same direction (see Fig 8.2), but when of unlike signs they tend to counteract each other. A $+P$ or $+cZ$ causes a $+B$, and a $-P$ or a $-cZ$ causes a $-B$.

Coefficient B is calculated by taking half the algebraic difference between the deviations on compass headings East and West:

$$B = \frac{\text{Deviation on E} - \text{Deviation on W}}{2}$$

Since components P and cZ cause deviation which varies as the sine of the aircraft heading, then deviation due to coefficient B may also be expressed as

$$\text{Deviation} = B \times \sin(\text{heading})$$

Coefficient C

Coefficient C represents the resultant deviation due to the presence, either together or separately, of hard-iron component Q and soft-iron component fZ (see Fig 8.3). These components when of like and unlike signs cause deviations whose directions are the same as those caused by components P and cZ. Therefore, a $+Q$ or a $+fZ$ causes a $+C$, and a $-Q$ or a $-fZ$ causes a $-C$.

Coefficient C is calculated by taking half the algebraic difference between the deviations on compass headings North and South:

$$C = \frac{\text{Deviation on N} - \text{Deviation on S}}{2}$$

Since components Q and fZ cause deviation which varies as the cosine of the aircraft heading then deviation due to coefficient C may also be expressed as

$$\text{Deviation} = C \times \cos(\text{heading})$$

Coefficient D

This coefficient represents the deviation due to the presence, either together or separately, of components aX and eY. It will be seen from Fig 8.7 (*a*) that these components cause deviation of the same type when they are of unlike signs and counteract each other when of like signs. When a $+aX$ or $-eY$ predominates, or when they are present together, coefficient D is said to be positive, whilst $-aX$ or $+eY$ predominating or together cause a negative coefficient D.

Coefficient D is one-quarter of the algebraic difference between the sum of the deviations on compass headings North-East and South-West, and the sum of the deviations on headings South-East and North-West, i.e.

$$D = \frac{(\text{Deviation on NE} + \text{Deviation on SW}) - (\text{Deviation on SE} + \text{Deviation on NW})}{4}$$

The deviations caused by components aX and eY vary as the sine of twice the aircraft heading; therefore deviation may also be expressed as

Deviation $= D \times \sin$ (twice heading)

Coefficient E

Coefficient E represents the deviation due to the presence of components bY and dX of like signs. When a $+bY$ and a $+dX$ are combined, coefficient E is said to be positive, whilst a combination of a $-bY$ and a $-dX$ give a negative coefficient; the two components must in each case be equal.

Coefficient E is calculated by taking one-quarter of the algebraic difference between the sum of the deviations on compass headings North and South, and the sum of the deviations on headings East and West:

$$E = \frac{(\text{Deviation on N} + \text{Deviation on S}) - (\text{Deviation on E} + \text{Deviation on W})}{4}$$

The deviations caused by components bY and dX vary as the cosine of twice the aircraft heading; therefore deviation may also be expressed as

Deviation $= E \times \cos$ (twice heading)

The total deviation on an uncorrected compass for any given direction of the aircraft's heading by compass may be expressed by the equation

Total deviation $= A + B\sin\theta + C\cos\theta + D\sin 2\theta + E\cos 2\theta$

ADJUSTMENT AND DEVIATION COMPENSATION OF COMPASSES

In order to determine by what amount compass readings are affected by hard- and soft-iron magnetism, a special calibration procedure known as *swinging* is carried out so that adjustments can be made and the deviation compensated.

DEVIATION COMPENSATION DEVICES

These devices fall into two distinct groups, mechanical and electromagnetic, the former being employed with simple direct-reading compasses and detector elements of certain remote-indicating compasses, and the latter being designed solely for use with detector elements of remote-indicating compasses.

In both cases, the function is the same, i.e. to neutralize the effects of the components of an aircraft's hard- and soft-iron magnetism by setting up opposing magnetic fields.

Mechanical Compensation Devices

One of the earliest mechanical devices is the micro-adjuster shown in Figs 8.11 and 8.12. It consists of two pairs of magnets (a feature common to all types of mechanical compensator), each pair being fitted in bevel gears made of a non-magnetic material. The gears are mounted one above the other so that, in the neutral condition, one pair of magnets lies fore-and-aft for the correction of coefficient C, and the other pair lies athwartships for the correction of coefficient

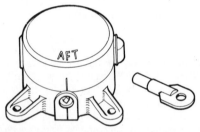

Fig 8.11 Micro-adjuster type of deviation compensator

B. Production of magnetic fields required for correction is obtained by rotating small bevel pinions which mesh with the gears, causing them to rotate in opposite directions. As can be seen from Fig 8.12, the magnets are thus made to open up in the manner of a pair of scissors, the fields being produced between the poles and in a direction dependent on that in which the operating head is rotated.

Let us now refer to Fig 8.13 in order to see what effects the fields produced by each pair of magnets have on a compass detector system. At (a) it is assumed that the aircraft is positioned on a Northerly heading, and due to an athwartships component (Q and its allied soft iron) of aircraft magnetism, the detector system is deviated a certain number of degrees from North, e.g. in an Easterly direction. To eliminate this deviation the *athwartships operating head* must be rotated in a direction which will open up the *fore-and-aft magnets* and create an athwartships field sufficient to deflect the compass magnet system Westerly and back to North.

Similar compensation for deviation occurring on a Southerly heading would also be effective as in (c). On Easterly or Westerly headings, however, the fore-and-aft magnets are ineffective because their fields are then aligned with the compass magnet system and so cannot cause deflection (Fig 8.13 (b) and (d)).

From the series of diagrams relating to Component P and athwartships corrector magnets (Fig 8.13 (e) to (h)) it will be noted that their fore-and-aft fields are effective on Easterly or Westerly headings.

Thus, for each pair of magnets compensation is effective only on two cardinal headings; namely fore-and-aft magnets, North and South; athwartships magnets, East and West.

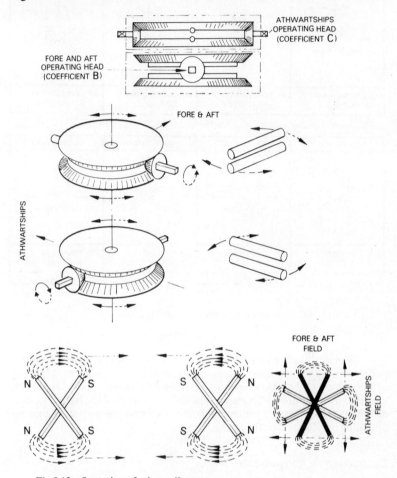

Fig 8.12 Operation of micro-adjuster

Two other versions of mechanical compensator are shown in Fig 8.14, the one at (*a*) being employed with the E2 series of standby compass (see p. 164), and the one at (*b*) with certain flux-detector units. Although differing in size and construction, they both employ gear-operated corrector magnet arrangements, the magnets being positioned side by side and not one above the other as in the micro-adjuster.

The magnets are mounted on flat gears meshing with each other and connected to operating heads, which in the compensator shown at (*a*), are operated by a key, and in that at (*b*) by means of screwdriver.

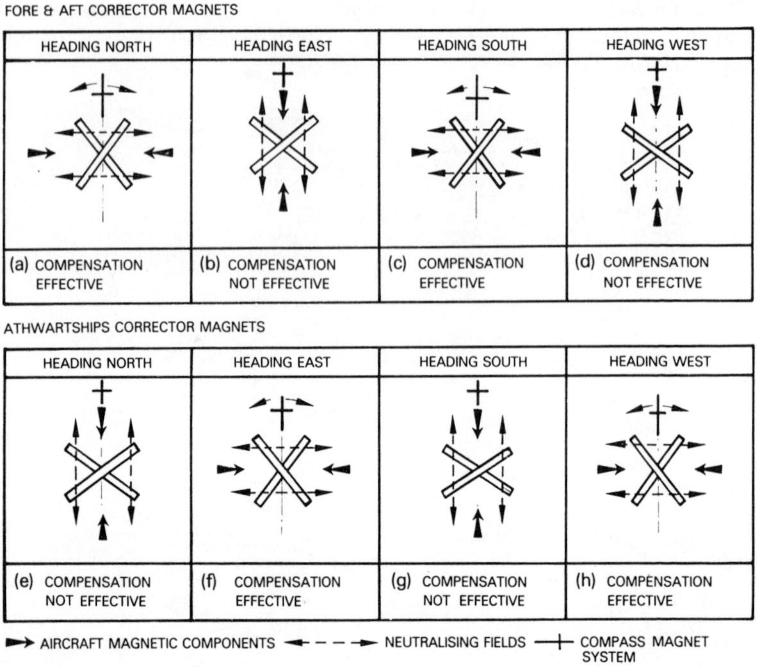

218   *Aircraft Instruments*

FORE & AFT CORRECTOR MAGNETS

HEADING NORTH	HEADING EAST	HEADING SOUTH	HEADING WEST
(a) COMPENSATION EFFECTIVE	(b) COMPENSATION NOT EFFECTIVE	(c) COMPENSATION EFFECTIVE	(d) COMPENSATION NOT EFFECTIVE

ATHWARTSHIPS CORRECTOR MAGNETS

HEADING NORTH	HEADING EAST	HEADING SOUTH	HEADING WEST
(e) COMPENSATION NOT EFFECTIVE	(f) COMPENSATION EFFECTIVE	(g) COMPENSATION NOT EFFECTIVE	(h) COMPENSATION EFFECTIVE

→ AIRCRAFT MAGNETIC COMPONENTS   ← − − → NEUTRALISING FIELDS   —|— COMPASS MAGNET SYSTEM

Fig 8.13   Effects of fields produced by magnets

As the magnets are rotated, their fields combine to set up neutralizing components in exactly the same manner as those of a micro-adjuster. Maximum compensation of deviation on either side of a quadrantal heading is obtained when the magnets are in complete alignment. Indication of the neutral position of the magnets is given by aligning datum marks on the ends of the magnet operating spindles and on the casing.

Electromagnetic Compensation Devices

From the foregoing descriptions of mechanical compensation devices, it is obvious that the required deviation compensations are carried out at the location of the compass or flux-detector element. This is a fairly common procedure but in certain remote-indicating compass systems means are provided which permit electromagnetic compensation of magnetic effects at flux detector elements to be carried out from within the aircraft. The main components of a typical electromagnetic compensator are shown in Fig 8.15.

The coil unit, which is mounted directly above the flux detector, is made up of four coils wound on Mumetal cores, one pair being positioned on the fore-and-aft line and the second pair athwartships. They thus correspond to the pairs of permanent magnets of mechanical compensators.

Aircraft Magnetism and its Effect on Compasses 219

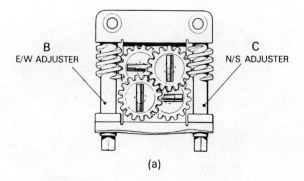

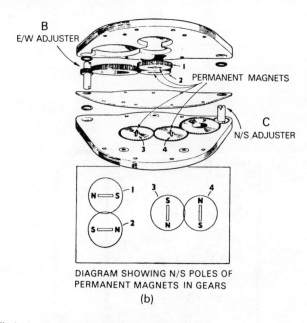

Fig 8.14 Mechanical compensation devices
(a) Direct-reading compass   (b) Flux detector element

The corrector unit contains two potentiometers which are supplied with direct current at approximately 8. V. Referring to the circuit diagram (Fig 8.16), it will be noted that each potentiometer is tapped at its mid-point and connected to one end of its corresponding pair of coils. The other ends of the coils are connected to wipers which can be rotated over the potentiometers, either side of the centre tap, by means of adjusting spindles protruding through the front panel of the corrector unit. Corrections for deviations up to $\pm 15°$ are thus obtainable, as indicated on plastic scales around the adjusting spindles.

220 Aircraft Instruments

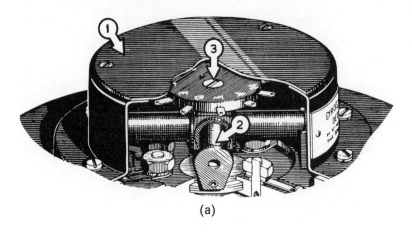

(a)

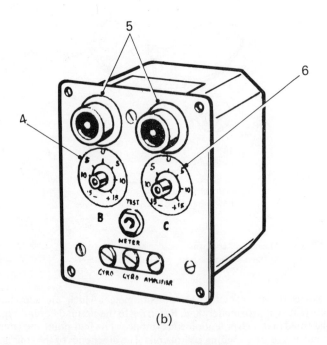

(b)

Fig 8.15 Components of an electromagnetic compensator
(a) Compensator   (b) Corrector control box
1 Compensator   4 E/W adjuster
2 Corrector coils   5 Control lamps
3 Terminal disc   6 N/S adjuster

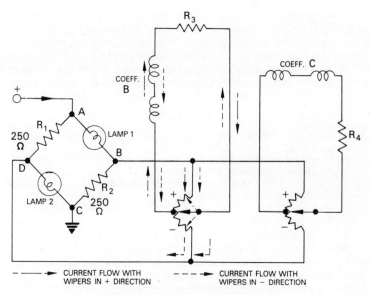

Fig 8.16   Circuit diagram of an electromagnetic compensator

An essential requirement of electromagnetic compensation devices is that, once neutralizing fields have been set up in the coils, they must remain constant during the time the electrical power supply is on and the compass system is in operation. It is therefore necessary to compensate for the effects of any fluctuations in the power supply. The method by which this requirement is met is governed by the type and design of compass system, but the one shown in Fig 8.16 illustrates the fundamental principle. It consists of a bridge network, connected across the main d.c. power supply and the two potentiometers, and is made up of two resistors and two special lamps. The filaments of the lamps, which are supplied as matched pairs, are made from a material having a high resistance/temperature coefficient; therefore the effective resistance of each lamp will depend upon its filament temperature.

When the supply voltage increases, the current also increases, and if this were not corrected it would increase the magnetizing effect of the coils. However, by including the lamps in the circuit, the increase of current through the filaments also increases their temperature and resistance; thus the current increase is immediately opposed. Conversely, a decrease in supply voltage and current will result in a decrease of filament resistance, the supply current being brought up to its nominal value. The current through the compensation circuit is of the order of 111–131 mA.

Under normal operating conditions, the resistance of the lamps is approximately 125 Ω, which is about half that of the current-limiting resistors R_1 and R_2. If the positive side of the d.c. supply is applied to the bridge junction A, the voltage drop across AB being smaller than that across BC, junction B will be at a higher potential than junction D with respect to ground. This, therefore, establishes the positive ends of the potentiometers—the top ends in the diagram.

With the potentiometer wipers at the zero, centre-tap position, no current can flow in any of the compensator coils and so they remain magnetically inert. By moving the wipers into the positive or negative sections of the potentiometers, current is caused to flow in the direction indicated, and a neutralizing magnetic field of the relevant polarity is obtained in each pair of coils. The resistors R_3 and R_4 in series with the compensator coils provide additional current limiting.

QUESTIONS

8.1 In connection with the effects of component Q which of the following statements is true?
  (a) They correspond to those of a magnet lying fore and aft and produce maximum deviation on North or South.
  (b) They correspond to a magnet lying athwartships and produce maximum deviation on North and South.
  (c) They correspond to a magnet lying athwartships and produce maximum deviation on East and West.

8.2 Under what attitude conditions of an aircraft will component R produce the same effects as positive and negative components P and Q?

8.3 (a) How are the nine soft-iron components of magnetism designated, and on which axes do their induced fields lie?
  (b) Which soft-iron components have the same effects as components P and Q?

8.4 Explain how the flux detector element of a remote-indicating compass system can be affected by components of aircraft magnetism.

8.5 (a) What are coefficients A, B, C, D and E?
  (b) Name the components of magnetism associated with each coefficient.
  (c) What is the difference between coefficients real A and apparent A?

8.6 (a) Express the formulae used for the calculation of coefficients A, B and C.
  (b) Given the following information find values for each of the three coefficients:

Magnetic heading	Compass deviation	Magnetic heading	Compass deviation
000°	+4°	180°	−1°
045°	+2°	225°	−2°
090°	+4°	270°	−2°
135°	+3°	315°	0°

8.7 An aircraft has a component $-P$ and a component $+Q$.
  (a) Draw separate curves of deviation caused by them.
  (b) What effect would the $-P$ have on the compass when heading 360° and the $+Q$ when heading 270°?

8.8 What do you understand by compass "swinging"?

8.9 With the aid of diagrams explain how a deviation compensating device neutralizes the fields due to aircraft magnetic components.

8.10 Describe the construction and operation of a typical electromagnetic compensating device.

9 Accelerometers and Fatigue Meters

The structure of an aircraft is designed to withstand certain stresses which may be imposed on it during flight, the magnitude of such stresses being dependent on the forces acting on the aircraft. All these forces may be resolved into components acting in the directions of the three mutually perpendicular axes of the aircraft, namely the longitudinal or roll axis, the lateral or pitch axis, and the vertical or yaw axis.

Force is the product of mass and acceleration, and since the mass of an aircraft may be considered constant, the forces acting on the aircraft in flight may be expressed in terms of the accelerations affecting it. During the manoeuvring of an aircraft, the largest changes in the accelerations to which it is subjected take place in the direction of the vertical axis. Consequently, the danger of exceeding allowable stresses and the possibility of failure of some part of the structure are greatest when excessive accelerations are applied through the vertical axis. Vertical components of acceleration are measured by means of instruments known as *accelerometers* and *fatigue meters*, and instruments of either sort may be installed in an aircraft. They operate on the same basic principle, but whereas the accelerometer provides instantaneous indications of vertical acceleration, the fatigue meter is designed to count and indicate the number of times predetermined threshold values of vertical acceleration have been exceeded.

BASIC ACCELEROMETER PRINCIPLE

A basic mechanism is illustrated in Fig 9.1, from which it will be noted that the principal components are a mass and two calibrated springs tensioned to statically balance the weight of the mass. The position of the mechanism shown in the diagram is the one corresponding to normal straight and level flight.

In this condition, the force along the vertical axis is due to the aircraft's weight, and therefore the aircraft is subject to normal gravitational force, which produces an acceleration of $1g = 32\,\text{ft/sec}^2$. Similarly, the force on the mass is due to its weight and so it too is subject to $1g$. The weight of the mass extends one spring and allows the other to contract. The pointer, which is actuated by the lever arm on which the weight is mounted, moves over a scale calibrated directly in g units, and for the level flight condition it is positioned at what may be termed the datum value of $1g$.

It will be noted from the diagram that the scale is calibrated in positive and negative values of g; these correspond respectively to upwards and downwards accelerations along the vertical axis. Thus, the load supported by the wings of an aircraft in any manoeuvre is the product of the aircraft's weight and the accelerometer indication.

Under vertical acceleration conditions brought about by manoeuvring of the aircraft, gusts or turbulent air, the mass will be displaced thus changing the tension of the springs until it balances the force imposed and producing the corresponding change in indication. A positive acceleration moves the mass downwards and a negative acceleration moves it upwards.

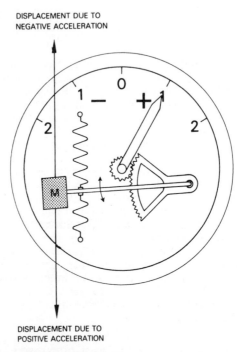

Fig 9.1 Basic accelerator mechanism

Typical Accelerometer

The dial presentation and schematic arrangement of the mechanism of a typical accelerometer are shown in Fig 9.2.

The mechanism consists of two spring-controlled masses mounted on cantilever arms attached to two rocking shafts. A sector gear attached to one of these shafts meshes with a pinion on the instantaneous pointer spindle. At the rear end of the second rocking shaft a sector gear meshes with a pinion coupled to a magnetic eddy-current drag device. The purpose of this is to damp out violent pointer fluctuations which could occur under short-period accelerations.

As the accelerometer scale is linear, there must be a linear relationship between acceleration and angular movement of the masses. This is achieved by using linear springs and attaching them to links secured to the rocking shafts at right angles to the cantilever arms. The rotation of the masses is therefore directly proportional to the extension of the springs, and hence to the acceleration force imposed.

When the aircraft and instrument are subjected to a vertical acceleration, each mass moves through an arc causing the rocking shafts to move in opposite directions, thus driving the instantaneous pointer to indicate the *g* force imposed.

In this particular design of accelerometer, it is also possible for horizontal components of acceleration which may act in the plane of the instrument dial

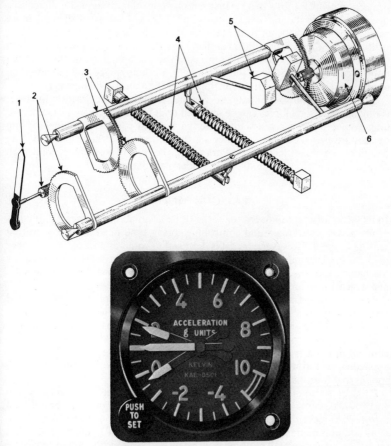

Fig 9.2 Typical accelerometer
1 Instantaneous *g* pointer
2 Pointer drive
3 Compensating gears
4 Control springs
5 Masses
6 Damping device

to exert forces on the two masses causing them to rotate the rocking shafts to false positions. However, since the rotation of each shaft would be in the same direction, any tendency to rotate at all can be prevented by gearing the shafts together. This is precisely the function of the two sector gears at the forward end of each shaft.

In addition to knowing the instantaneous value of acceleration, it is also very important to know the maximum acceleration experienced during a manoeuvre or when flying in turbulent conditions. It is customary therefore to provide accelerometers with two auxiliary pointers, one to indicate maximum positive accelerations and the other to indicate maximum negative accelerations.

Both pointers are mounted concentrically with the "instantaneous g" pointer and are driven by a small plate attached to this pointer. The plate has projections which move two ratchet wheels; engaging in opposite directions, one wheel drives the maximum positive pointer, and the other the maximum negative pointer, to the position of the main pointer. A lightly-loaded pawl and spring hold the auxiliary pointers at their respective maximum indications until that acceleration is exceeded, or until reset by the operation of the resetting knob located at the front of the instrument. Both auxiliary pointers are then returned to the position of the main pointer by the action of hairsprings.

Since it is the purpose of an accelerometer to indicate acceleration forces, it is obvious that such forces can be imposed on it in the course of general handling. To avoid damage which might be sustained by careless handling, locking devices are provided which may be set from the rear of the instrument. The design of the actual mechanism varies with the type of instrument, but in all cases the locking function is to prevent movement of the masses.

FATIGUE METER

In its basic form, a fatigue meter is the same as an accelerometer, i.e. it comprises a suspended mass and controlling-spring type of vertical-acceleration sensing element. However, since it is designed to count and record the number of times predetermined acceleration threshold values have been exceeded, its mechanism is of necessity a little more complex. An external view and schematic of a typical meter are illustrated in Figs 9.3 and 9.4 respectively.

The mass of the acceleration sensing element consists of a weight mounted on a cantilever spring which is connected by secondary springs and a fusee chain to a rotary inertia and damping unit. The inertia unit is, in turn, connected by a shaft to a wiper arm and brush assembly which sweep around the face of a commutator. Each segment of the commutator represents an equal change of applied acceleration, e.g., $0 \cdot 1g$. The centreportion of the commutator corresponds to the normal gravitational force of $1g$.

Pairs of commutator segments are connected electrically to relays which energize electromagnetic digital counter units corresponding to the selected threshold values, in this case $0 \cdot 05g$, $0 \cdot 45g$, $0 \cdot 75g$, $1 \cdot 25g$, $1 \cdot 55g$ and $1 \cdot 95g$. Each relay circuit is made up of two sections referred to as "lock" and "release," and from the schematic it will be noted that the "lock" sections are connected to the segments which also correspond to the selected threshold values. The "release" sections are connected to segments corresponding to lower values which are a definite number of increments apart. For example, the relay controlling the counter recording $0 \cdot 75g$ has its "lock" section connected to segment $0 \cdot 75$ and its "release" section connected to segment $1 \cdot 05$, an incremental difference of $0 \cdot 30g$. It should also be noted that, in the positive and negative directions, the connections for the "release" sections are taken from segments which precede those connected to the "lock" sections.

The power for operating the relays and counters is 24 V direct current. The instrument is rigidly attached to the aircraft structure in accessible underfloor compartments or special compartments within a passenger cabin.

Fig 9.3 Fatigue meter—external view

Operation

As in an indicating accelerometer, the mass can be moved from the normal $1g$ position by the forces acting on it. Assuming that the aircraft and meter are subjected to a vertical acceleration in the positive direction, then the mass will move downwards and will cause the wiper arm to move in a clockwise direction through an angular distance proportional to the applied acceleration.

When the wiper brushes reach the first commutator segment connected to the "lock" section of a relay and associated counter, a circuit is completed through relay coil (1). The relay contacts are thus closed and connect a positive supply to the counter solenoid which operates and cocks the counter; no reading is recorded at this moment. In addition to completing a circuit to the counter solenoid, the relay contacts also feed a positive supply to a hold-in coil (2) of the relay. If the acceleration is of such a value that the mass and wiper arm

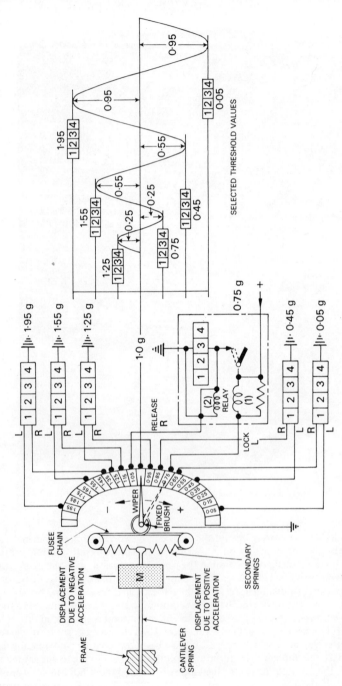

Fig 9.4 Fatigue meter—schematic arrangement

cause the brushes to make contact with the succeeding selected segments in the positive direction, then their associated relays will be energized in a similar manner and the counters cocked. Thus the hold-in coils keep the relays energized even though the brushes leave their particular commutator segment.

When the acceleration forces begin to decay to the $1g$ value, the brushes return over the commutator, making contact with segments connected to the "release" sections of the relays. Reference to the diagram shows that when a relay is still energized its contact also feeds a positive supply to the segment connected to its "release" circuit segment. Therefore, when the brushes reach the segment a circuit is completed which shorts out the relay hold-in coil, causing the contact to open. The counter solenoid is thus de-energized and thus releases a spring-loaded pawl of the cocking mechanism to complete the counting operation.

The reason for arranging the circuit in this manner is to prevent a second count being made when the brushes move to a high g level and return over a selected segment. Furthermore, it also prevents extra counts when accelerations of small amplitude and high frequency are superimposed on the gust or manoeuvre acceleration.

Excessive overshooting of the mass is prevented by eddy-current drag induced by the rotation of a bell-shaped damper in a permanent-magnetic field.

As fatigue meters are only required to count and record the vertical accelerations experienced in flight, it is necessary to include in the electrical circuit an automatic means of switching the power supply on and off. This is accomplished by an airspeed switch unit (see page 95) whose contacts are set to make the circuit just after take-off and to break it just before landing.

The acceleration counts recorded by a fatigue meter are essential for relating the accelerations experienced by an aircraft to what is termed its safe fatigue life. In consequence, readings have to be noted at regular specified periods and entered on special tabulated data sheets which on completion must be used for analysis and calculation of fatigue life.

QUESTIONS

9.1 With the aid of a diagram explain the basic operating principle of an accelerometer.
9.2 What are the essential differences between an accelerometer and a fatigue meter?
9.3 How is it ensured that a fatigue meter does not record (*a*) second "counts," (*b*) during taxying and landing?

10 Synchronous Data-transmission Systems

With the introduction of large multi-engined aircraft the problem arose of how to measure various quantities such as pressure, temperature, engine speed and fuel tank contents at points located at greater distances from the cockpit. Many of the instruments then available operated on mechanical principles which could be adapted to suitably transmit the required information. For example, on one very early twin-engined aircraft, mechanically-operated engine speed indicators were designed with large-diameter dials so that by mounting the indicators in the engine nacelles they could be read from the cockpit.

However, as aircraft were further increased in size and complexity, the adaptation of mechanically-operated instruments became severely limited in application. A demand for improved methods of measurement at remote points therefore arose and was met by changing to the use of electrical systems in which an element detects changes in the measured quantity and transmits the information electrically to an indicating element.

We can therefore consider most of the instruments used in a present-day aircraft as being of the remote-indicating type, but many of them are of a design in which the transmission of data is effected through the medium of a special synchronous system.

Synchronous systems fall into two classes: direct-current and alternating-current. The principles of some of those commonly used form the subject of this chapter. Although varying in the method of data transmission, all the systems have one common feature: they consist of a transmitter located at the source of measurement and a receiver which is used to position the indicating element.

DIRECT-CURRENT SYNCHRONOUS SYSTEMS

The Desynn System

This system, one of the earliest to be used in aircraft, may take one of three forms, namely: *rotary motion* or *toroidal resistance* for position and liquid-contents indications; *linear motion* or *micro-Desynn* for pressure indication, and *slab-Desynn* also for pressure indication. The principle of operation is the same in each case, but the rotary-motion arrangement may be considered as the basic system from which the others have been developed.

The Basic System
The transmitting and receiving components of this system are shown in Fig 10.1. The electrical element of the transmitter consists of a resistor wound on a circular former (called the "toroidal resistor") and tapped at three points 120° apart. Two diametrically-opposed wiper contact arms, one positive and the other negative, are insulated from each other by a slotted arm which engages with a pin actuated by the appropriate mechanical element of the transmitter.

The wiper contact arms are assembled in the form of a bar having rotational freedom about a pivot which carries current to the positive arm. Current to the negative arm is carried via a wiper boss the underside of which is in contact with a ring fitted on the inner side of the terminal moulding. A circlip, fitted at the end of the pivot, holds the complete assembly in place against a spring which gives the required contact pressure on the toroidal resistor.

The receiver element consists of a cylindrical two-pole permanent-magnet rotor pivoted to rotate within the field of a laminated soft-iron stator, carrying a star-connected three-phase distributed winding supplied from the toroidal resistor tappings. A tabular brass housing is fitted inside the stator, and together

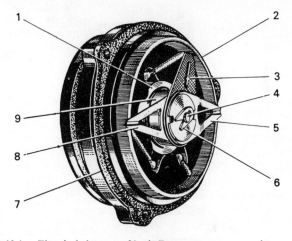

Fig 10.1A Electrical elements of basic Desynn system: transmitter

1 Contact ring (negative)
2 Toroidal resistor
3 Slotted arm
4 Pivot (positive)
5 Contact arm (positive)
6 Circlip
7 Terminal moulding
8 Contact arm (negative)
9 Wiper base

with its end cover, provides a jewelled bearing support for the rotor spindle. The front end of the spindle projects through the end cover and a dial mounting plate, to carry the pointer. Electrical connection between the transmitting and receiving elements may be either by terminal screws or plug-type connectors.

The electrical elements of the receivers are common to all three circuit arrangements of the Desynn system.

Operation

When direct current is applied to the transmitter contact arms, which are in contact with the toroidal resistor, currents flow in the resistor causing the three tapping points to be at different potentials. For example, with the contact arms in the position shown in Fig 10.2, the potential at tapping No 2 is greater than that at No 1 because there is less resistance in the circuit between the positive arm and the No 2 tapping. Thus, currents are caused to flow in the lines between transmitter and receiver, the magnitude and direction of which depend upon the position of the contact arms on the toroidal resistor.

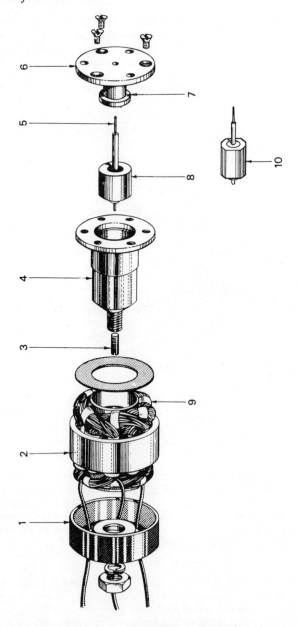

Fig 10.1b Electrical elements of basic Desynn system: indicator

1 End cover
2 Soft-iron ring
3 Bearing screw
4 Rotor housing
5 Pointer shaft
6 End-plate
7 No-voltage magnet
8 Cylindrical rotor
9 Stator
10 Octogonal rotor

In turn, these currents flow through the coils of the receiver stator and produce a magnetic field about each coil similar to that of a bar magnet; thus either end of a coil may be designated as a N pole or a S pole, depending on the direction of the current through a coil. The combined fields extend across the stator gap and cause the permanent-magnet rotor to align itself with their resultant.

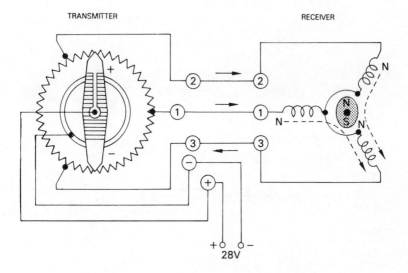

Fig 10.2 Circuit diagram of basic Desynn system

Figure 10.3A shows how the potentials at the transmitter tapping points are changed when the contact arms are rotated, and also how the resultant magnetic field at the receiver follows this rotation. For purposes of calibration the zero position of a transmitter is defined by coincidence of the positive contact with tapping No 3.

The potentials produced for a 360° rotation of the contact arms are also shown, and from the graph in Fig 10.3B it will be noted that they follow a sawtooth waveform. Now, the instantaneous sum of three such waves mutually displaced by 120° is not always zero and as a result the magnetic field of the stator does not rotate in perfect synchronism with the contact arms, being subject to what is termed a cyclic error that reaches a maximum every 60°. However, since this deviation characteristic is a constant, it can be taken into account during the initial calibration of the indicating element, its only disadvantage being to introduce a slight non-linearity of the scale.

The directions of the magnetic fields in the receiver stator depend on the relative values of current and not absolute values; therefore, the supply voltage to a Desynn system is not critical.

It will be noted from the diagram of the receiver element (Fig 10.1) that a pull-off magnet is fitted to the end plate. The purpose of this magnet is to act as

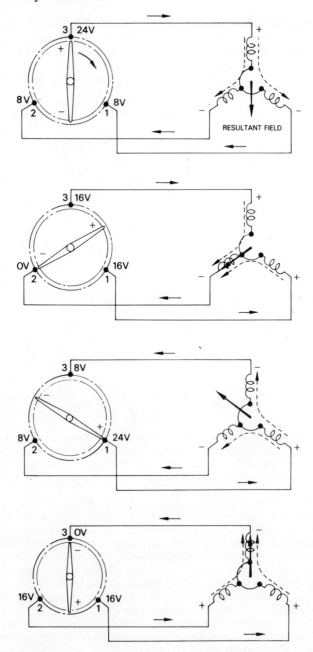

Fig 10.3A Potentials at toroidal transmitter tapping points

a power-failure device by exerting an attractive force on the main magnet rotor so as to pull it and the pointer to an off-scale position when current to the stator is interrupted. The strength of this magnet is such that, when the system is in operation, it does not distort the main controlling field.

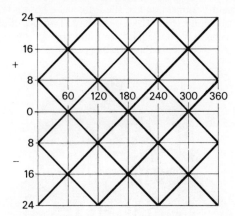

Fig 10.3B   Transmitter waveform

The Micro-Desynn System

In applications where the movement of a prime mover is small and linear, the use of a basic-system transmitting element is strictly limited. The micro-Desynn transmitter shown in Fig 10.4 was therefore developed to permit the magnification of such small movements and to produce, by linear movement of contacts, the same electrical results as the complete rotation of the contact arms of the basic transmitter.

In order to understand the development of this transmitting element, let us imagine that a toroidal resistor has been cut in two, laid out flat with its ends joined together, and three tappings made as before together with positive and negative arms in contact with it. Movement of the contact arms will produce varying potentials at the tappings, but as will be clear from Fig 10.5(a), the full range will not be covered because one or other of the arms would run off the resistor. We thus need a second resistor with tappings so arranged that the contact arms can move through equal distances.

If we now take two toroidal resistors and join them in parallel, then by cutting them both in two and laying them out flat, we obtain the circuit arrangement shown at (b). By linking the contact arms together and insulating them from each other, they can now be moved over the whole length of each resistor to produce voltage and current combinations which will rotate the receiver rotor through 360°. Since the contact arms have to traverse a much shorter path, their angular movement can be kept small (usually 45°), a feature which helps to reduce the energy required to operate the transmitter.

The resistors are wound on bobbins which may be of round or square section, the latter type being designed to help reduce cyclic and friction errors.

Each resistance bobbin is secured in place against a set of miniature spring contact fingers accurately positioned so as to provide the necessary tapping points.

The contact arms are mounted on a rocker shaft supported between the vertical parts of a U-shaped bracket, and movement of the transmitter's mechanical element is transmitted to the arms via a spring-loaded operating pin and crank arm connected to the rocker shaft. Two beryllium-copper hairsprings conduct current to the contact arms and also act together to return the rocker shaft and contact arms to their starting position.

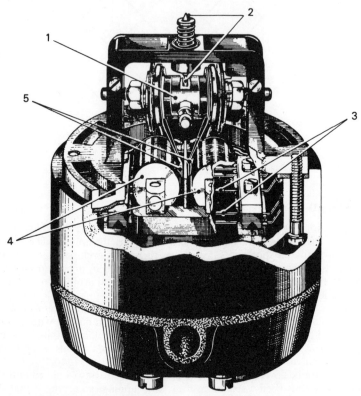

Fig 10.4 Typical micro-Desynn transmitter element

1 Pivoted rocker shaft
2 Spring-loaded operating pin and crank arm
3 Fixed contact fingers
4 Coils
5 Contacts

The Slab-Desynn System

In addition to the cyclic error present in the basic and micro-type systems, small errors also arise due to friction set up by the contact arms having to move over a considerable surface of resistance wire. Although such errors can be reduced by providing a good contact material and by burnishing the resistance wire surface, the cyclic error is still undesirable in certain measurements.

A solution to this problem was brought about by modifying the basic system so as to change the three sawtooth waveforms into sinusoidal waves, the instantaneous sum of which is always zero. The transmitter so developed is shown schematically in Fig 10.6, from which it will be noted that the resistor and contact arms have, as far as electrical connections are concerned, virtually changed places with each other. The resistor is now wound on a slab former,

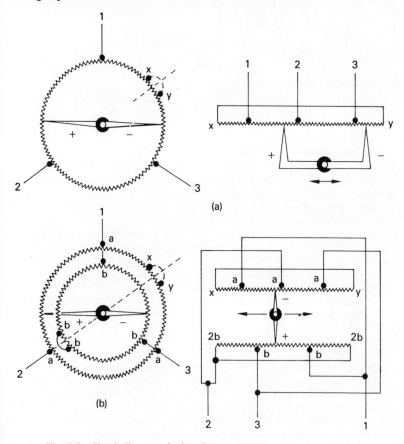

Fig 10.5 Circuit diagram of micro-Desynn system

hence the term "slab-Desynn," and is connected to the direct-current supply, while the contact arms themselves provide the three potential tapping points for the indicator stator.

The three contact arms are insulated from each other and pivoted over the centre of the slab, and are each connected to a slip ring. Spring-finger brushes bear against these slip rings and convey the output currents to the stator coils. Movement of the mechanical element is transmitted to the brushes via a gearing system.

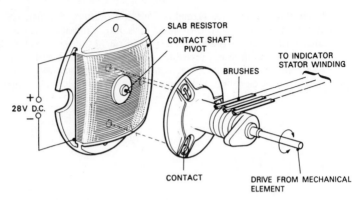

Fig 10.6 Slab-Desynn transmitter

Step-by-step Transmission System

This type of transmission system is direct-current operated and was originally developed by the Admiralty for use as a repeating system in conjunction with marine gyro-compasses. Its application to aircraft instrumentation is limited to certain types of remote-indicating compass for the azimuth monitoring of an autopilot and associated navigational equipment.

The transmitter is in the form of a cylindrical commutator arranged in two portions, one connected to the negative side and the other to the positive side of the direct-current supply. Both portions are insulated from each other, and we may consider it as a double-pole rotary switch. At two diametrically-opposite points on the circumference there is an insulated or neutral segment subtending an angle of 30°. Five carbon brushes are disposed around the commutator; two of them are connected to the direct-current supply and so positioned that both portions of the commutator always carry a potential. The three other brushes are positioned at 120° to each other and connected via line wires to the receiver.

When the commutator is rotated, the energization being fed to the stator coils of the receiver will change every 30°. If we imagine the commutator as being laid out flat as shown in Fig 10.7, then it is clear that for one complete revolution of the commutator the polarities and stator energization will change twelve times as indicated in the table. The rotating magnetic field set up produces a torque on the receiver rotor tending to pull its magnetic axis into line with the field, and thus causing it to take twelve successive "steps" per revolution; hence the name "step-by-step transmission." It is of interest to note that, although the system depends on direct current for its operation, the commutator segment and brush arrangement produces a rotating magnetic field similar to that of the stator of a three-phase alternating-current motor.

A disadvantage of this type of transmission system is that it is not self-aligning and requires lining up before use. Also, in the event of a power failure, the receivers connected to the transmitter get out of line. Therefore, conventional lining-up positions are adopted for a transmitter, namely when it is feeding line 3 negatively and lines 1 and 2 positively, or alternatively, line 3 positively and lines 1 and 2 negatively. Either alternative represents the same position of the magnetic flux in the receiver although the direction of the flux is reversed. Reference to the table (Fig 10.7) shows that a lining-up position occurs twice in every twelve steps, i.e. on steps 1 and 7.

Each pair of stator windings is connected between two line wires and so the back-e.m.f. (back electromotive force) of self-induction in the coils will be carried via the line wires back to the transmitter, and will be apparent in the form of a spark between brushes and commutator. To reduce this effect, suppressors are connected in delta form across the lines.

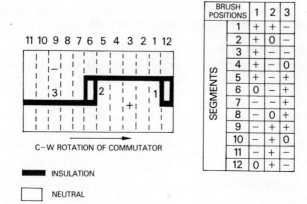

Fig 10.7 Step-by-step transmission system

ALTERNATING-CURRENT SYNCHRONOUS SYSTEMS

Systems requiring alternating current for their operation are usually classified under the generic term *synchro* and are manufactured under various contracted trade names, e.g. Autosyn, Selsyn, Magnesyn. All these systems, with the exception of the Magnesyn, operate on the same principle and may be considered as special forms of rotary transformer. Since the Magnesyn is somewhat of an outsider its operating principle will be considered separately at a later stage.

Before going into the operating principles of synchros, it is necessary to have a clear understanding firstly of the process of electromagnetic induction and secondly its application to a standard type of transformer.

Electromagnetic Induction

Electromagnetic induction refers broadly to the production of an e.m.f. within a conductor when there is relative movement between it and a magnetic field.

If a conductor is moved in a magnetic field the lines of force will be cut and an e.m.f. will be induced in the conductor of a magnitude proportional to the rate at which the lines are cut. If the ends of the conductor are connected together, thus forming a closed circuit, a current resulting from the e.m.f. will flow so long as the conductor is moving.

The effect would be the same if the conductor were stationary and the field were moving. When current is passed through a conductor and is increased or decreased, the field will proportionately increase or decrease, causing the flux lines around each turn of the coil to move in an expanding and contracting manner thus cutting adjacent turns and inducing an e.m.f. in the coil. If the current is constant no e.m.f. will be induced.

The property described in the foregoing paragraphs is known as *self-inductance* and is the fundamental principle on which motors and generators operate.

Another inductive effect we may now consider is that when two current-carrying coils are placed in close proximity to each other they become linked by the fluxes they produce, so that, when the current through one coil is changed, an e.m.f. will be induced not only in that coil but also in the adjacent coil. This property is known as *mutual inductance* and is utilized in the transformer.

Principle of the Transformer

A transformer, shown in Fig 10.8, employs two electrically separate coils on an iron core. One of the coils, the primary, is connected to an alternating-voltage supply and the other, known as the secondary, is provided with terminals from which an output voltage is taken. The alternating-voltage applied to the primary causes an alternating current to flow in this winding, the current being dependent on the inductance of the winding. The effect of the alternating current is to set up an alternating magnetic flux in the core, and since most of this flux links

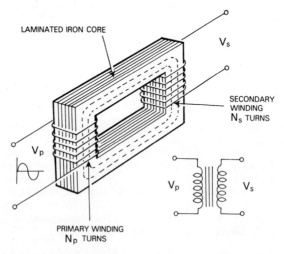

Fig 10.8 Principle of transformer

with the secondary winding, an alternating e.m.f. is produced in the secondary by mutual induction. If the secondary terminals are connected to a closed circuit, a current will flow and all the energy expended in the circuit is transferred magnetically through the core from the supply source connected to the primary terminals. As already stated, the primary and secondary windings are on a common core, which means that the magnetic flux within the core will, in addition to linking the secondary winding, also link the primary winding. Thus an e.m.f. is set up in the primary winding which, neglecting resistance, can be considered for practical purposes as being equal and opposite to the voltage applied to the primary. If the primary winding has N_p turns the voltage per turn is V_p/N_p. Since the same flux cuts the secondary it will induce in it the same voltage per turn and the secondary voltage will thus be

$$V_s = \frac{V_p}{N_p} N_s$$

so that

$$\frac{\text{Primary voltage}}{\text{Secondary voltage}} = \frac{V_p}{V_s} = \frac{N_p}{N_s}$$

The ratio of voltages in the two windings is therefore proportional to the ratio of the turns, known as the *turns ratio* of the transformer.

There is one further effect concerning secondary current which should be considered. When the current flows it too sets up a flux which opposes the main core flux. A reduction in the core flux reduces the primary e.m.f. opposing the applied voltage, and so allows increased current to flow in the primary winding. This increased current then restores the core flux to a value which is only very slightly less than its no-load value.

Neglecting transformer losses, which are generally very small, the input power may be equated to the output power:

$$V_p I_p = V_s I_s$$

so that

$$\frac{I_p}{I_s} = \frac{V_s}{V_p} = \frac{N_s}{N_p}$$

from which it is clear that the currents are inversely proportional to the numbers of turns.

Basic Synchro System

As a starting-point in the understanding of a synchro system let us consider the basic Desynn system and see how it performs when the transmitter is connected to an alternating-current supply as shown in Fig 10.9 (*a*). The alternating supply will establish a magnetic field in the receiver stator the position of which will be governed by the position of the contact arms. However, for any position of the contact arms, the field is continually reversing polarity, and for the frequencies of the alternating current normally used, the permanent magnet in the receiver cannot reverse its direction so rapidly. It will respond only to the average torque created by a coil, and since this is zero the magnet will assume some position independent of the action of the stator coils.

We can, of course, substitute for the permanent magnet a coil wound on a core and by connecting it to the supply in the manner shown at (*b*) obtain a magnet the polarity of which will reverse at the same time as those of the coils. This arrangement still does not provide us with a true synchronous link between transmitter and receiver; for, although the electromagnet is able to follow the stator fields, it can still assume independent positions and get out of step.

Let us now consider the arrangement of Fig 10.10 (a) from which is will be noted that the transmitter is made up of a stator and rotor identical to those of the receiver. The stator coils are connected at points S_1, S_2 and S_3 in a similar manner to those of a three-phase generator and motor system, while the rotors are single-phase connected at points R_1 and R_2 to the alternating-current

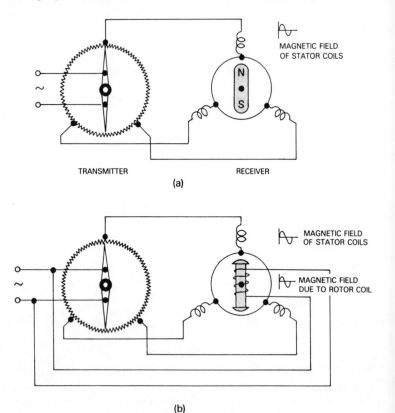

Fig 10.9 Development of an a.c. synchro system
(a) Basic Desynn system connected to an a.c. supply
(b) Effect of a rotor coil within the receiver

supply. The similarity between this arrangement and a transformer will also be noted; the rotors correspond to primary windings and the stators to secondary windings.

When the rotors are aligned with their respective stators in the position indicated they are said to be at "electrical zero." This refers to the angle standardized for synchros at which a given set of stator voltages will be produced, and enables replacement synchros to be matched to each other. Other positions are measured in degrees increasing in an anticlockwise direction when viewed from the output-shaft end of the unit.

Synchronous Data-transmission Systems 243

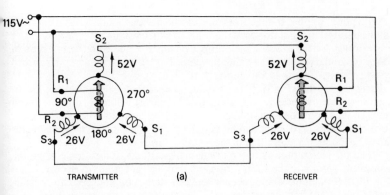

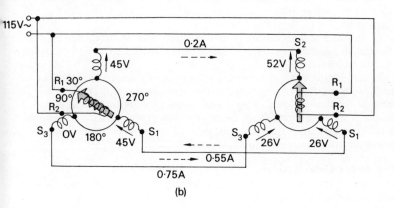

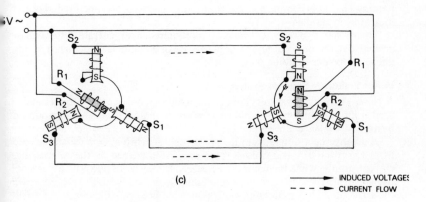

Fig 10.10 A.C. synchro transmitter and receiver circuits
 (a) Rotors aligned (b) Transmitter rotor at 30° (c) Fields produced

With power applied to the rotors, due to transformer action a certain voltage will be induced in the stator coils the value of which will be governed, as in any transformer, by the ratio of the number of turns of the rotor (primary) and stator (secondary) coils. The values given in Fig 10.10 are based on a synchro system operated from a 115 V supply.

Referring again to Fig 10.10 (*a*), we can now see that, with power applied to the transmitter and receiver rotors, equal and opposite voltages will be produced between their interconnecting lines when the rotors occupy the same angular positions. Thus, no current can flow in the stator coils and no magnetic fields can be produced to cause either of the rotors to turn.

When the rotors occupy different angular positions, for example when the transmitter rotor is at the 30° position and the receiver rotor is at electrical zero (Fig 10.10 (*b*)), an unbalance occurs between stator coil voltages causing current to flow in the lines and stator coils. The currents are greatest in the circuits where voltage unbalance is greatest, and their effect is to produce magnetic fields which exert torques to turn the receiver rotor to the same position as that of the transmitter (Fig 10.10 (*c*)).

As the receiver rotor continues to turn, the misalignment, voltage unbalance and currents decrease until the 30° position is reached and no further torque is exerted on the rotor.

In considering this synchronizing action one might assume that, since currents are also flowing in the stator coils of the transmitter, its rotor would line up again at electrical zero. This is a reasonable assumption, because in fact a torque is set up tending to turn the rotor in a clockwise direction. However, it must be remembered that the rotor is being actuated by some prime mover which exerts loads too great to be overcome by the rotor torques.

Repeater or Data Synchros

In some applications of synchros, it is necessary to repeat the information or data presented by a receiver at several stations; e.g. when applied to a compass system required for use in an aircraft having a navigator's station, heading information must be available at this station as well as at the pilot's instrument panels. This is accomplished by connecting two synchro-operated repeater indicators in parallel.

If more than two repeater synchros are required, a larger transmitter will be needed to maintain accuracy, or, as in the case of the compass system referred to above, a repeater amplifier unit.

Power Follow-up Synchro System

A *power follow-up* synchro system is one which produces error voltages and is used in applications where output torques greater than those normally produced by a synchro receiver are required.

Its arrangement differs from that of a basic synchro system as can be seen from Fig 10.11. The transmitter rotor only is connected to the alternating-current supply, while the rotor of the receiver, which is known as a "control transformer," is used as an inductive winding for detecting the direction at which the error signal is a minimum. Furthermore, it is electrically connected to an amplifier and mechanically connected to a two-phase motor. Another difference to be noted is that the control-transformer rotor is circular in shape and is at electrical zero when positioned at right angles to the transmitter rotor, so that no

Synchronous Data-transmission Systems 245

voltage is induced in its coil. The stator coils of a control transformer are wound with more turns of wire than in a normal synchro; the purpose of this being to make the impedance high enough to prevent high currents from flowing.

When the position of the transmitter rotor relative to its stator is changed, a voltage is induced in the control-transformer stator, and if its rotor is not aligned with the resultant magnetic field an error voltage signal is induced in the rotor coil. The signal, the amplitude of which is a measure of the rotor displacement, is amplified and fed to the control phase of the motor, the other phase being continuously energized from the alternating-current supply. With both phases energized the motor will rotate, driving the load connected to it and also

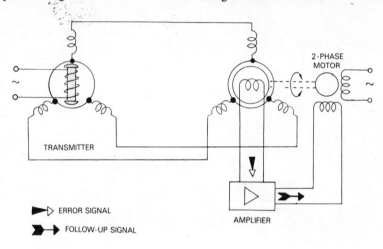

Fig 10.11 Power follow-up synchro system

the control-transformer rotor, so reducing its displacement relative to the transmitter rotor. The rotation continues until the rotor is aligned with the resultant magnetic field at which position no further error voltage is induced, and the load has been moved by an amount proportional to the angular displacement of the transmitter rotor.

Resolver Synchro

In addition to the fundamental synchronizing functions studied so far, it is necessary in some applications for input signals to be resolved into precise angular measurements, and their sine and cosine components. For such measurements yet another type of synchro is required, namely the *resolver synchro*.

In its physical appearance, a resolver resembles a normal synchro, but from the circuit arrangement shown in Fig 10.12 it will be noted that the stator and rotor each consist of two windings 90° apart; *phase quadrature*, as it is called. If the rotor is turned through an angle θ and a signal voltage V is applied to the rotor windings, the components $V \sin \theta$ and $V \cos \theta$ can be found with the aid of tables or by drawing a right-angled triangle as shown. The output signals will be $kV \sin \theta$ and $kV \cos \theta$, where k is the transformation ratio.

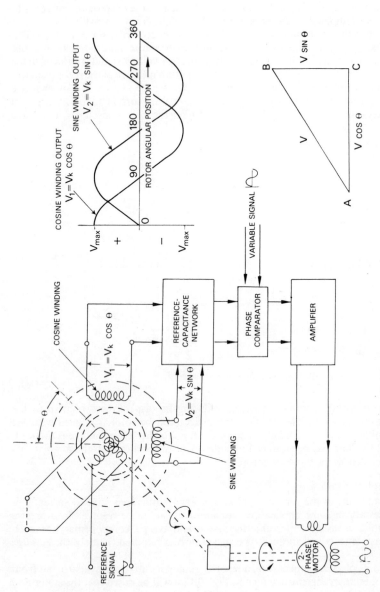

Fig 10.12 Principle of a resolver synchro

Reference signal voltages are applied to one winding of the resolver rotor; the other winding is usually short-circuited as this improves accuracy. The alternating magnetic field associated with the energized winding induces voltages V_1 and V_2 in the cosine and sine windings of the stator, respectively. Now, as in a normal synchro the voltages are dependent upon the angular position of the rotor, so that, when the energized winding is parallel to one of the stator windings, then maximum voltage will be induced in this winding, and when at right angles minimum voltage will be induced. As the stator windings are in phase quadrature, the output signals are resolved components of induced voltage proportional to the cosine and sine respectively of the rotor's angular position.

The stator voltages are fed to a resistance-capacitance network and due to its action the phasing of the output is related to the position of the resolver rotor, i.e. in phase quadrature. The voltages produced by the network, when added vectorially, thus combine to produce an output that changes in phase, degree for degree of rotor movement.

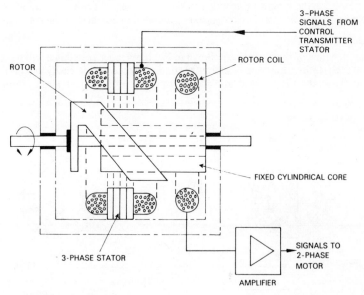

Fig 10.13   Synchrotel

Synchrotel

A *synchrotel* is generally used as a low-torque control transformer or transmitter. It employs a conventional 3-phase stator, but as will be noted from Fig 10.13, unlike a conventional synchro the rotor section is in three separate parts: a hollow aluminium cylindrical rotor of oblique section, a fixed single-phase rotor winding, and a cylindrical core about which the rotor rotates. The rotor shaft is supported in jewelled bearings and is connected to the pressure-sensing element or whatever element the application demands.

Aircraft Instruments

In a typical pressure-measuring application, the synchrotel is electrically connected to a synchro control transmitter whose rotor is made to follow the synchrotel rotor position; in other words, it acts as a servo loop system.

The synchro transmitter rotor is energized by a 26 V 400 Hz single-phase supply which induces voltages in the transmitter stator. As this stator is connected to the synchrotel stator then a resultant radial alternating flux is established across it. For any particular pressure applied to the sensing element, there will be a corresponding position of the synchrotel rotor, and due to its oblique shape, sections of it will be cut by the radial stator flux. Currents are thus produced in the rotor, and since it is pivoted around the cylindrical core, an axial component of flux will be created in the core. The rotor winding is also fixed around the core, and therefore the core flux will induce an alternating voltage in the winding, and the amplitude of this voltage will be a sinusoidal function of the relative positions of the rotor and stator flux. This voltage is fed, via an amplifier, to the control phase of a two-phase servo motor which drives the synchro transmitter rotor round in its stator, thereby causing a change in the synchrotel stator flux, to the point where no voltage is induced in the rotor winding, i.e. the synchro transmitter is driven to the null position. This position corresponds to the pressure measured by the sensing unit at that instant.

Magnesyn System

The elements of this system are shown in Fig 10.14, and from this it will be seen that each stator is made up of a circular laminated core of soft, easily magnetized material such as Permalloy, and around this core is placed a single continuous

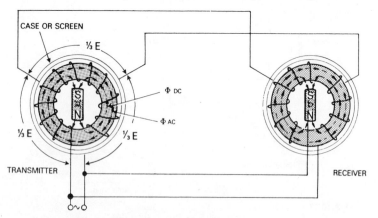

Fig 10.14 Circuit diagram of Magnesyn system

toroidal winding connected to a source of alternating current. Surrounding the toroid is an annular stack of core laminations which acts as a shield against extraneous magnetic fields that may be present near the units.

Each winding is divided by two taps into three equal sections, corresponding taps of the transmitter and receiver being connected together.

The rotor is a two-pole permanent magnet, magnetized along a line perpendicular to its axis of rotation, and mounted in bearings on an axis passing through the centre of the stator.

Operation

By first considering the magnetic circuit of the transmitter it will be seen that a magnetic flux Φ_{DC}, is set up in each half of the core by the permanent-magnet rotor; the rotor pole axis bisects the core. Assuming the core to be made of a homogeneous material, symmetrical in shape and concentric with the rotor, the reluctance of the two paths will be the same and Φ_{DC} will be equal in the two halves of the core.

When the toroidal winding of the transmitter is supplied with alternating current, this will produce an alternating flux Φ_{AC} which we will consider at a specific time to rotate clockwise in the core. With respect to Φ_{AC}, the unidirectional flux Φ_{DC} is positive in one half of the core and negative in the other half. Thus the total flux in one half of the coil will be

$$\Phi_1 = \Phi_{AC} + \Phi_{DC}$$

and in the other half,

$$\Phi_2 = \Phi_{AC} - \Phi_{DC}$$

The total fluxes give rise to flux densities B_1 and B_2 which are made up of two components: B_{AC}, which is common to both halves of the core and varies at the same frequency as Φ_{AC}; and an additional flux density B_x, which exists because of the flux Φ_{DC} produced by the rotor combining with Φ_{AC}. Thus

$$B_1 = B_{AC} + B_x \quad \text{and} \quad B_2 = B_{AC} - B_x$$

When Φ_{AC} is zero the flux density of the core is low and the greatest amount of flux Φ_{DC} will be passing through the core. As Φ_{AC} increases during the positive half of the cycle the flux density of the core will obviously increase and this means that the permeability becomes less. The unidirectional flux therefore gets "pushed" out of the core, causing it to cut across the coil. It will be clear that for a decrease in Φ_{AC} the reverse of these conditions will occur, the flux Φ_{DC} again entering the core.

As Φ_{AC} goes to maximum in the negative half-cycle the same effects on permeability and unidirectional flux Φ_{DC} will take place. Thus, by plotting the variations in Φ_{DC} for one complete cycle we obtain curve C of Fig 10.15 which represents the effective flux linking the coil. The shape of the curve is due to the saturation and hysteresis characteristics of the core material.

Now, because voltage is induced in a winding surrounding a piece of magnetic material proportional to the rate of change of flux in the material, we see that the variations in Φ_{DC} cutting across the stator coil will induce a voltage in the coil. Furthermore, it will be a second-harmonic voltage due to the fact that Φ_{DC} varies twice from maximum to minimum during one cycle of Φ_{AC}. At a given instant, the pattern of the second-harmonic voltage will have a direction which depends on the position of the rotor with respect to the stator.

Let us assume for the moment that the receiver stator coil has no excitation voltage supplied to it. Then it is apparent that because of the interconnection between transmitter and receiver coil the voltage pattern produced in the transmitter will be reproduced in the sections of the receiver stator, and this will also cause a symmetrical second-harmonic current to flow in the receiver stator. The voltage and current are indicated respectively by the curves D and E in Fig 10.15 (b).

250 *Aircraft Instruments*

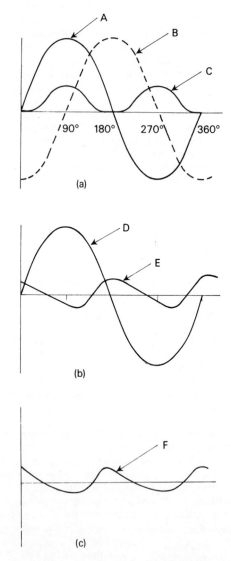

Fig 10.15 Flux and current produced in a Magnesyn system
(a) Transmitter (b) Receiver (c) Resultant flux

The second-harmonic current in turn produces a second-harmonic flux in the receiver stator core because there is no excitation voltage and therefore no excitation flux Φ_{AC} to change the permeability of the core. As the second-harmonic flux is alternating through the coil and across the core, it is obvious it will not produce accurate positioning of the receiver rotor with respect to that in the transmitter.

Synchronous Data-transmission Systems

Consequently, the flux must be made unidirectional by supplying it with excitation voltage. How this is accomplished may be explained as follows. The flux resulting from the excitation voltage increases the flux density of the core as it goes through both the positive and negative halves of the cycle. Now, and as may be seen from curves D and E of Fig 10.15 (b), the phase relationship between the excitation flux and the second-harmonic current is such that the flux in the receiver core is small during positive half-cycles of the second-harmonic current and large during negative half-cycles.

This means that, during the negative half-cycles of the second-harmonic current the excitation flux is so high that the core permeability becomes very low; as a result, the negative flux is low because of the greater reluctance of the path which this flux must follow. This produces a resultant flux shown as curve F in Fig 10.15 (c), which, having a positive average value, creates an effective unidirectional field across the receiver stator on an axis parallel to that existing across the transmitter.

The receiver rotor thus aligns itself with this field and duplicates the position of the transmitter rotor in its stator.

QUESTIONS

10.1 Draw a circuit diagram of the basic Desynn system and explain its operating principle.

10.2 Explain how the basic system is developed for such applications as pressure measurement.

10.3 (a) Is it essential for the power supply to a Desynn system to remain exactly at the designed operating value?
(b) How are the indicator pointers returned to the "off-scale" position?

10.4 Explain the difference between the following systems, giving examples of an aircraft application of each: (a) Desynn, (b) micro-Desynn, (c) slab-Desynn.
(SLAET)

10.5 (a) What is the difference between self-induction and mutual induction?
(b) Explain how each of these principles applies to an aircraft component with which you are familiar.

10.6 By means of a circuit diagram explain how the transformer principle is applied to a basic synchro system.

10.7 In what type of synchro system is a control transformer utilized? Explain the operating principle.

10.8 In what manner does a Synchrotel differ in construction and operation from the more conventional synchro systems?

10.9 Explain how a synchro is applied to systems involving the measurement of the sine and cosine components of angles.

10.10 Describe the construction of a Magnesyn unit and explain its operation when used as a data transmission system.

11 Measurement of Engine Speed

The measurement of engine speed is of considerable importance, since together with such parameters as manifold pressure, torque pressure and exhaust gas temperature, it permits an accurate control over the performance of the engine to be maintained.

With reciprocating engines the speed measured is that of the crankshaft, while with turboprop and turbojet engines the rotational speed of the compressor shaft is measured, such measurement serving as a useful indication of the thrust being produced. The indicating instruments are normally referred to as tachometers.

The method most commonly used for measuring these speeds is an electrical one, a method which long ago superseded those based on mechanical principles. However, some mechanical tachometers operating on the centrifugal governor principle are still in use and a review of their operating principles provides a useful starting-point for the subject.

CENTRIFUGAL TACHOMETER

This type of tachometer was the first ever to be used in aircraft for engine speed measurement, its operation being based on the principle of the centrifugal governor employed for controlling the speed of a stationary gas or steam engine.

As may be seen from Fig 11.1, a spindle carrying a pivoted weight is connected to the engine via a flexible drive shaft and a step-up gear train. The other end of the shaft is connected to a step-down gear train at the engine. One end of the weight is connected via a link to a muff free to slide up and down the spindle. A balance spring around the axis of the weight holds the weight at some angular position with respect to the driving spindle. The indicating element consists of a sector gear, which engages with the sliding muff, and a pinion fixed to a pointer spindle.

When the engine is running the flexible drive shaft is rotated at a lower speed than the engine, in order to reduce wear, and this in turn rotates the indicator spindle and weight via the step-up gear at the same speed as the engine. Due to centrifugal force acting on the weight the latter assumes a position towards horizontal, this motion being balanced by the spring, and causes the sliding muff to take up a certain position along the spindle. The movement of the muff is transmitted to the pointer, which takes up a position to indicate the balance between centrifugal force and spring tension in terms of revolutions per minute.

Flexible Drive Shafts

The construction of a typical flexible drive shaft is shown in Fig 11.2. The inner shaft embodies a central core of hardened steel wire over which five layers, each of four strands of finer gauge wire, are wound, alternate layers being wound in opposite directions. After the shaft has been cut to the appropriate length a connector is secured to each end by swaging. Both connectors incorporate shanks which engage hollow adapters provided at the engine drive shaft and in the indicator. The outer casing is a continuous winding of two specially formed steel wires and is arranged to be flexible, oiltight and waterproof. Flanged collars are swaged to each end of the casing to provide a means of attachment to the engine and indicator.

Axial movement of the inner shaft is restricted by a shoulder on the connector at the indicator end which abuts on the end of the flanged collar, and also by a slip washer.

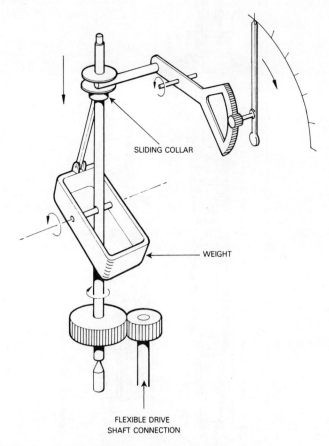

Fig 11.1 Centrifugal-type tachometer

ELECTRICAL TACHOMETER SYSTEMS

It will be apparent from the foregoing description of mechanical indicators that, because a flexible drive is necessary, there must be some restriction to their maximum length. This was a problem which had to be overcome with the introduction of multi-engined aeroplanes, the increasing distances between cockpits

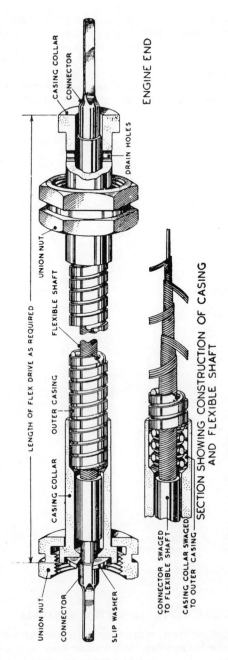

Fig 11.2 Construction of a flexible drive shaft

and engines making it very difficult to find suitable locations for drives and indicators. Electrical methods were therefore developed, the first taking the form of a small direct-current generator mounted near the engine and driven by a short flexible drive shaft. The generator voltage, proportional to engine speed, was transmitted to a voltmeter calibrated in revolutions per minute.

This method, however, also had its disadvantages, among which were the use of brushes and commutator, the necessity for individual adjustment of generators and indicators to produce the required calibration law, and the fact that the total possible error was the sum of both generator and indicator errors. The next stage in development was a changeover to an alternating-current system, which not only eliminates brushes but, in measuring speed in terms of frequency, has no errors introduced should the generator voltage vary over a wide range. Such systems having been developed to high standards of accuracy are now the accepted means of engine speed measurement in multi-engined aircraft.

Alternating-current System

The generator consists of a permanent-magnet rotor running within a slotted stator which carries a star-connected three-phase winding. The rotor may be either two-pole or four-pole, formed as shown in Fig 11.3, and is magnetized

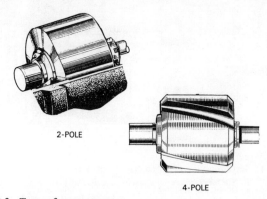

Fig 11.3 Types of generator rotor

after positioning within the stator. It will be noted that the poles of the four-pole rotor are angled so that when one end of a pole leaves one stator tooth the other end is entering the next tooth. This produces the best waveform and permits an even driving torque. With the two-pole rotor the same effect is achieved by skewing the stator teeth and the individual coils which make up a phase.

There are two main methods of driving the generator, one requiring the use of a short flexible drive shaft and generator gearbox, and the other in which the generator may be directly mounted on the engine or its associated accessories gearbox. The latter method, which usually takes the form of a splined shaft coupling, is employed on turbine engines where, due to higher driving speeds, the use of flexible drive shafts is impracticable.

In order to limit the mechanical loads on generators, the operational speed of rotors is generally fixed at a maximum of 5,000 rev/min. When used with piston engines, which have maximum crankshaft speeds below this figure, the generator rotor may be driven at the crankshft speed. Four-to-one or two-to-one ratio gears are then used between engine and generator to ensure lower rotational speeds and a reduction in wear of the flexible drive shafts.

The compressors of turbine engines run at much higher speeds (10,000 and 20,000 rev/min); therefore, to ensure that the rotors do not exceed maximum speed, the necessary speed reduction is effected by coupling the generators to a half-speed or quarter-speed shaft of the engine gearbox. A sectioned view of a typical generator is shown in Fig 11.4.

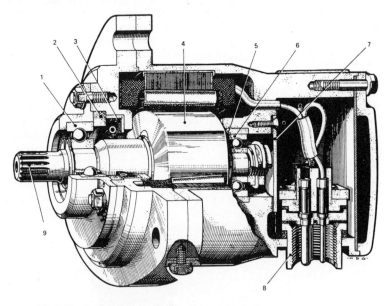

Fig 11.4   Sectioned view of spline-drive generator

1	Ball bearings	6	Ball bearings
2	Oil-seal retaining ring	7	Sealing cover
3	Oil seal	8	Breeze connector
4	Two-pole permanent-magnet rotor	9	Driving spline
5	Grease retainer		

A typical indicator, shown in Fig 11.5, consists of two interconnected elements: a driving element and an eddy-current-drag speed-indicating element.

Let us consider first the *driving element*. This is, in fact, a synchronous motor having a star-connected three-phase stator winding and a rotor revolving on two ball bearings. The rotor is of composite construction, embodying in one part soft-iron laminations, and in the other part a laminated two-pole permanent magnet. An aluminium disc separates the two parts, and a series of longitudinal copper bars pass through the rotor forming a squirrel-cage. The purpose of constructing the rotor in this manner is to combine the self-starting and high torque properties of a squirrel-cage motor with the self-synchronous properties associated with a permanent-magnet type of motor.

The *speed-indicating element* consists of a cylindrical permanent-magnet rotor inserted into a drum so that a small airgap is left between the periphery of the magnet and drum. A metal cup, called a *drag cup*, is mounted on a shaft and is supported in jewelled bearings so as to reduce frictional forces, in such a way that it fits over the magnet rotor to reduce the airgap to a minimum. A calibrated hairspring is attached at one end of the drag-cup shaft, and at the

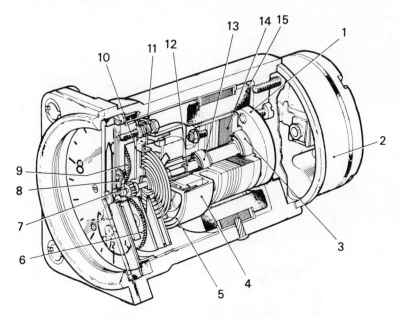

Fig 11.5 Sectioned view of a typical a.c.-type tachometer indicator

1 Cantilever shaft
2 Terminal block assembly
3 Rear ball bearing
4 Magnetic cup assembly
5 Drag element assembly
6 Small pointer spindle and gear
7 Outer spindle bearing
8 Bearing locking tag
9 Intermediate gear
10 Bearing plate
11 Hairspring anchor tag
12 Inner spindle bearing
13 Front ball bearing
14 Rotor
15 Stator

other end to the mechanism frame. At the front end of the drag-cup shaft a gear train is coupled to two concentrically mounted pointers; a small one indicating hundreds, and a large one indicating thousands, of revolutions per minute.

Operation

As the generator rotor is driven round inside its stator, the poles sweep past each stator winding in succession so that three waves or phases of alternating e.m.f. are generated, the waves being 120° apart (see Fig. 11.6). The magnitude of the e.m.f. induced by the magnet depends on the strength of the magnet and the number of turns on the phase coils. Furthermore, as each coil is passed by a pair of rotor poles, the induced e.m.f. completes one cycle at a frequency determined by the rotational speed of the rotor. Therefore, rotor speed and frequency

258 *Aircraft Instruments*

are directly proportional, and since the rotor is driven by the engine at some fixed ratio then the frequency of induced e.m.f. is a measure of the engine speed.

The generator e.m.f.s are supplied to the corresponding phase coils of the indicator stator to produce currents of a magnitude and direction dependent on the e.m.f.s. The distribution of stator currents produces a resultant magnetic field which rotates at a speed dependent on the generator frequency. As the

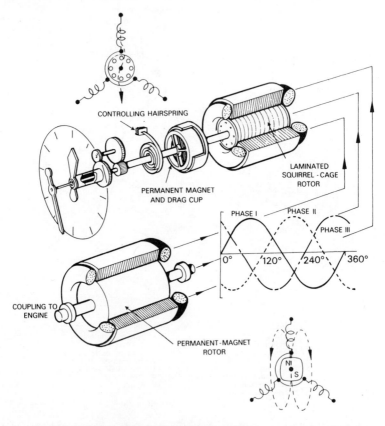

Fig 11.6 Principle of a.c.-type tachometer system

field rotates it cuts through the copper bars of the squirrel-cage rotor, inducing a current in them which, in turn, sets up a magnetic field around each bar. The reaction of these fields with the main rotating field produces a torque on the rotor causing it to rotate in the same direction as the main field and at the same speed.

As the rotor rotates it drives the permanent magnet of the speed-indicating unit, and because of relative motion between the magnet and the drag-cup, eddy currents are induced in the latter. These currents create a magnetic field which reacts with the permanent-magnetic field, and since there is always a tendency to oppose the creation of induced currents (Lenz's law), the torque reaction of the fields causes the drag-cup to be continuously rotated in the same direction as the magnet. However, this rotation of the drag-cup is restricted by the calibrated hairspring in such a manner that the cup will move to a position at which the eddy-current-drag torque is balanced by the tension of the spring. The resulting movement of the drag-cup shaft and gear train thus positions the pointers over the dial to indicate the engine speed prevailing at that instant.

Indicators are compensated for the effects of temperature on the permanent magnet of the speed-indicating element by fitting a thermo-magnetic shunt device adjacent to the magnet. It operates in a similar manner to the compensator described on page 21.

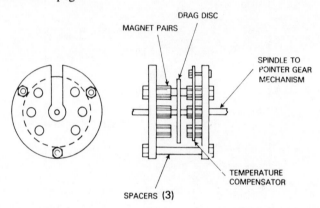

Fig 11.7   Disc-type of drag element

Figure 11.7 shows another version of speed-indicating element which is used in some types of indicator. It consists of six pairs of small permanent magnets mounted on plates bolted together in such a way that the magnets are directly opposite each other with a small airgap between pole faces to accommodate a drag disc. Rotation of the disc is transmitted to pointers in a similar manner to the drag-cup method.

Percentage Tachometer

The measurement of engine speed in terms of a percentage is now widely adopted, particularly for turbojet engine operation, and was introduced so that various types of engine could be operated on the same basis of comparison. Furthermore, the dial presentation, shown in Fig 11.8, improves the readability of the instrument. The main scale is calibrated from 0–100% in 10% increments, with 100% corresponding to the optimum turbine speed. In order to achieve this the engine manufacturer chooses a ratio between the actual turbine speed and the generator drive so that the optimum speed produces 4,200 rev/min at the generator drive. A second pointer moves over a subsidiary scale calibrated in 1% increments.

A percentage tachometer system operates on the same principle and consists of the same basic elements as the normal alternating-current system. There are, however, certain essential constructional differences between generators which should be noted.

The generator is much smaller than the normal type and consists of a two-pole permanent-magnet driven by a square-ended drive shaft provided with a spring quill to allow slight misalignment of the generator on its mounting to be taken up. The generator is lubricated by circulating oil through it from the engine oil system, the casing of the generator being sealed to withstand a pressure of 40 lb/in^2.

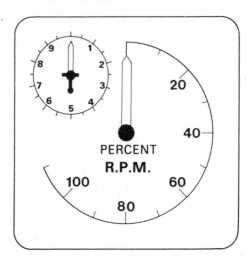

Fig 11.8   Dial presentation of a percentage tachometer indicator

SYNCHROSCOPES

In aircraft powered by combinations of reciprocating, turboprop, and turbojet engines, the problem arises of maintaining the engine speeds in synchronism at "on-speed" conditions and so minimizing the effects of structural vibration and noise.

The simplest method of maintaining synchronism between engines would be to manually adjust the throttle and speed control systems of the engines until the relevant tachometer indicators read the same. This, however, is not very practical for the simple reason that individual instruments can have different permissible indication errors; therefore, when made to read the same operating speeds, the engines would in fact be running at speeds differing by the indication errors. In addition, the synchronizing of engines by a direct comparison of tachometer indicator readings is made somewhat difficult by the sensitivity of the instruments causing a pilot or engineer to overshoot or undershoot an on-speed condition by having to "chase the pointers."

In order to facilitate manual adjustment of speed an additional instrument known as a *synchroscope* was introduced. It provides a qualitative indication of the differences in speeds between two or more engines, and by using the technique of setting up the required on-speed conditions on a selected master engine, the instrument also provides a clear and unmistakeable indication of whether a slave engine is running faster or slower than the master.

The instrument was designed at the outset for operation from the alternating current generated by the tachometer system, and it therefore forms an electrical part of this system. The dial presentations of synchroscopes designed for use in twin and four-engined aircraft are shown in Figs 11.9 (*a*) and (*b*) respectively, while a combination dual r.p.m. and synchroscope presentation is shown at (*c*).

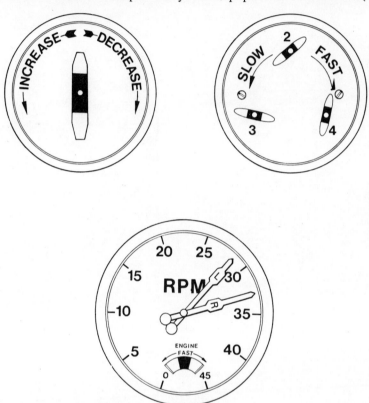

Fig 11.9   Dial presentations of synchroscopes
(*a*) Twin-engine
(*b*) Four-engine
(*c*) Combined dual tachometer and synchroscope

The operation is based on the principle of the induction motor, which, for this application, consists of a three-phase star-wound laminated stator and a three-phase star-wound laminated rotor pivoted in jewelled bearings within the stator. The stator phases are connected to the r.p.m. generator of the slave engine while the rotor phases are connected to the master engine generator via slip rings and wire brushes. A disc at the front end of the rotor shaft provides for balancing of the rotor. The pointer, which is double-ended to symbolize a propeller, is attached to the front end of the rotor shaft and can be rotated over a dial marked INCREASE at its left-hand side and DECREASE at its right-hand side.

On some synchroscopes the left-hand and right-hand sides may be marked

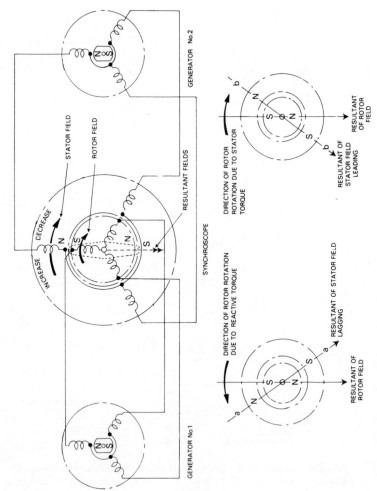

Fig 11.10 Operation of a synchroscope

SLOW and FAST respectively. Synchroscopes designed for use in four-engined aircraft employ three separate induction motors, the rotor of each being connected to the master engine tachometer generator while each stator is connected to one of the three other generators.

Operation

To understand the operation of a synchroscope let us consider the installation of a typical twin-engined aircraft tachometer system, the circuit of which is given in Fig 11.10. Furthermore, let us assume that the master engine, and this is usually the No 1, has been adjusted to the required "on-speed" condition and that the slave engine has been brought into synchronism with it.

Now, both generators are producing a three-phase alternating-current for the operation of their respective indicators, and this is also being fed to the synchroscope, generator No 1 feeding the rotor and No 2 the stator. Thus, a magnetic field is set up in the rotor and stator, each field rotating at a frequency proportional to its corresponding generator frequency, and for the phase rotation of the system, rotating in the same direction. For the conditions assumed, and because generator frequencies are proportional to speed, it is clear that the frequency of the synchroscope stator field is the same as that of the rotor field. This means that both fields reach their maximum strength at the same instant; the torques due to these fields are in balance, and the attraction between opposite poles keeps the rotor "locked" in some stationary position, thus indicating synchronism between engine speeds.

Consider now the effect of the slave engine running slower than the master. The frequency of the slave engine generator will be lower than the master engine generator, and consequently the stator field will be lagging behind the rotor field; in other words, reaching its maximum strength at a later instant at, say, point *a* in Fig 11.10. The rotor, in being magnetized faster than the stator, tries to rotate the stator and bring the stator field into alignment, but the stator is a fixed unit; therefore, a reactive torque is set up by the interaction of the greater rotor torque with the stator. This torque causes the rotor to turn in a direction opposite to that of its field so that it is forced to continually realign itself with the lagging stator field. The continuous rotation of the rotor drives the propeller-shaped pointer round to indicate that the slave engine is running SLOW and that an INCREASE of speed is required to bring it into synchronism with the master engine.

If the slave engine should run faster than the master then the synchroscope stator field would lead the rotor field, reaching maximum strength at, say, point *b*. The stator field would then produce the greater torque, which would drive the rotor to realign itself with the leading stator field, the pointer indicating that the slave engine is running FAST and that a DECREASE of speed is required to synchronize it.

As the speed of the slave engine is brought into synchronism once again, the generator frequency is changed so that a balance between fields and torques is once more restored and the synchroscope rotor and pointer take up a stationary position.

From the foregoing description we see that a synchroscope is, in reality, a frequency meter, its action being due only to the relative frequencies of two or more generators. The generator voltages play no part in synchroscope action except to determine the operating range above and below synchronism.

ROTATION INDICATORS

During the starting cycle of by-pass turbine engines, severe damage may arise if the shaft of the low-pressure compressor and turbine assembly is not free to rotate. Indicators are therefore provided as part of the instrumentation for these engines, to indicate commencement of rotation and that it is safe to continue the starting cycle.

The basis of an indicator is a two-stage magnetic amplifier operating from a 115 V 400 Hz a.c. supply and connected to one phase of a normal tachometer generator. The signals from the generator are fed into the amplifier as a reference input in revolutions per minute. An indicator lamp mounted on the main instrument panel, or flight engineer's panel, is connected to the amplifier output stage.

When the speed/voltage input signal reaches a critical level, usually 6 mV corresponding to a rotation speed of a fraction of 1 rev/min, sufficient output current is produced to light the indicator lamp. The speed is reached in the first few degrees of rotation; therefore the lamp provides an immediate indication that the low-pressure shaft has rotated. Signals in excess of the critical cause the amplifier to saturate and the lamp to remain alight but without being overloaded.

The power supply is fed to the amplifier via an engine-starting circuit and is isolated once the starting cycle is satisfactorily concluded. In multi-engine installations, a single amplifier and lamp serve to indicate rotation of each engine, being automatically selected during each starting cycle.

QUESTIONS

11.1 Describe a method of providing indication of engine speed. (SLAET)

11.2 Explain how rotation of an alternating-current tachometer indicator motor is transmitted to the indicating element.

11.3 With the aid of a circuit diagram explain the operation of a synchroscope system as fitted to a multi-engined aircraft. (SLAET)

11.4 (a) How is the necessary speed reduction between engine and tachometer generator obtained when direct-drive generators are employed?
(b) Why are direct-drive generators essential for the measurement of turbine engine speeds?

11.5 How are the effects of temperature on an alternating-current tachometer indicator compensated?

11.6 Why do you think it is unnecessary for indicators and generators to be used in matched pairs?

12 Measurement of Temperature

METHODS AND APPLICATIONS

In most forms of temperature measurement, the variation of some property of a substance with temperature is utilized. These variations may be summarized as follows:

1. Most substances expand as their temperature rises; thus, a measure of temperature is obtainable by taking equal amounts of expansion to indicate equal increments of temperature.
2. Many liquids, when subjected to a temperature rise, experience such motion of their molecules that there is a change of state from liquid to vapour. Equal increments of temperature may therefore be indicated by measuring equal increments of the pressure of the vapour.
3. Substances change their electrical resistance when subjected to varying temperatures, so that a measure of temperature is obtainable by taking equal increments of resistance to indicate equal increments of temperature.
4. Dissimilar metals when joined at their ends produce an electromotive force (*thermo-e.m.f*) dependent on the difference in temperature between the junctions. Since equal increments of temperature are only required at one junction, a measure of the electromotive force produced will be a measure of the junction temperature.

The utilization of these variations provides us with a very convenient method of classifying temperature-measuring instruments: (*a*) *expansion type* (liquid or solid), (*b*) *vapour-pressure type*, and (*c*) *electrical type* (resistance or thermoelectric).

The majority of aircraft instruments are of the resistance and thermoelectric types and are applied to the temperature measurement of liquids and gases such as fuel, engine lubricating oil, outside air, carburettor air, and turbine exhaust gas. Liquid-expansion and vapour-pressure thermometers, which pioneered the field of temperature measurement in aircraft, are nowadays strictly limited in their applications and are to be found only in a few types of light aircraft.

HEAT AND TEMPERATURE

Heat

Heat is a form of energy possessed by a substance, and is associated with the motion of the molecules of that substance. The hotter the substance the more vigorous is the vibration and motion of its molecules. We may regard it, therefore, as molecular energy.

The quantity of heat a substance contains is dependent upon its temperature, mass and nature of the material from which the body is made. A bucketful of warm water will melt more ice than a cupful of boiling water; the former must therefore contain a greater quantity of heat even though it is at a lower temperature.

The transfer of heat from one substance to another may take place by conduction, convection and radiation. *Conduction* requires a material medium, which may be solid liquid or gaseous. Hot and cold substances in contact interchange heat by conduction. *Convection* is the transmission of heat from one place to another by circulating currents and can only occur in liquids and gases. *Radiation* is the energy emitted by all substances, whether solid, liquid or gaseous.

Temperature

Temperature is a measure of the "hotness" or "coldness" of a substance or the quality of heat. Therefore, in the strictest sense of the term, temperature cannot be measured. The temperatures of substances can only be compared with each other and the differences observed, and so the practical measurement of temperature is really the comparison of temperature differences. In order to make such a comparison, the selection of a standard temperature difference, a fundamental interval and an instrument to compare other temperature differences with this, are necessary.

Melting Point and Boiling Point

For pure substances, the change of state from solid to liquid and from liquid to vapour takes place at temperatures which, under the same pressure conditions, can always be reproduced. Thus, there are two equilibrium temperatures known as (i) *melting point*, the temperature at which solid and liquid can exist together in equilibrium, and (ii) *boiling point*, the temperature at which liquid and vapour can exist together in equilibrium.

Fundamental Interval and Fixed Points

The fundamental interval is the temperature interval or range between two fixed points: the *ice point* at which equilibrium exists between ice and vapour-saturated air at a pressure of 760 mmHg, and the *steam point*, at which equilibrium exists between liquid water and its vapour; the water boiling also under a pressure of 760 mmHg.

The term *fundamental interval* is used in resistance thermometry; it refers to the increase in resistance of the temperature-sensing element between the fixed points.

Scales of Temperature

The fundamental interval is divided into a number of equal parts or degrees the division being in accordance with two scale notations, Celsius (centigrade) and Fahrenheit.

On the Celsius scale, the fundamental interval is divided into 100 degrees, the ice point being taken as 0°C and the steam point as 100°C. One Celsius degree is thus 1/100 of the fundamental interval.

On the Fahrenheit scale, the fundamental interval is divided into 180 degrees, the ice point and steam point in this case being taken as 32°F and 212°F respectively. One Fahrenheit degree is thus 1/180 of the fundamental interval.

If all the heat were removed from a body its temperature would be as low as possible. This temperature is $-273\cdot15°C$, or $273\cdot15°C$ below freezing point, and is known as *absolute zero* or 0 on the *Kelvin** scale:

$$0\,K = -273\cdot15°C \qquad 0°C = 273\cdot15\,K$$

On the Fahrenheit scale, absolute zero is $-459\cdot67°F$, or $0°R$ on the *Rankine* scale:

$$0°R = -459\cdot67°F = 0\,K = -273\cdot15°C$$
$$32°F = 491\cdot67°R = 273\cdot15\,K = 0°C$$

Conversion Factors

Since 100 divisions on the Celsius scale correspond to 180 divisions on the Fahrenheit scale,

$$1°C = \frac{180}{100} = 1\cdot8°F \qquad \text{and} \qquad 1°F = \frac{1}{1\cdot8} = 0\cdot55°C$$

Expressed as fractions,

$$1°C = \frac{9}{5}°F \qquad \text{and} \qquad 1°F = \frac{5}{9}°C$$

The fact that the zero points on the two scales do not coincide makes conversion from one scale to another more difficult than just dividing by the conversion factors, $1\cdot8$ and $0\cdot55$. For example, if we wish to convert $10°C$ to $°F$ and simply multiply it by $1\cdot8$ or its equivalent 9/5, we shall obtain $18°F$, but this is only $18°$ above freezing point and the value required must be with reference to the zero on the Fahrenheit scale. We therefore add $32°$ to obtain $+50°F$. Thus the formula for converting $°C$ to $°F$ is

$$°F = (°C \times 1\cdot8) + 32 \qquad \text{or} \qquad °F = \left(°C \times \frac{9}{5}\right) + 32$$

To convert $°F$ to $°C$ is simply the reverse procedure. From $50°F$ we subtract 32 to determine the number of Fahrenheit degrees above freezing point ($18°$) and divide by $1\cdot8$ or 9/5. Thus, the formula for converting $°F$ to $°C$ is

$$°C = \frac{(°F - 32)}{1\cdot8} \qquad \text{or} \qquad °C = (°F - 32) \times \frac{5}{9}$$

When converting values on the minus side of scales care must be taken to observe the signs. For example, in converting $-10°C$ to $°F$ we obtain $(-18) + (+32)$ giving us a difference of $+14$, the equivalent of $-10°C$ in $°F$.

To convert $°C$ and $°F$ to degrees absolute we simply add 273 or 460 respectively.

LIQUID-EXPANSION AND VAPOUR-PRESSURE TEMPERATURE INDICATORS

The oldest of temperature-measuring instruments utilizing the principle of liquid expansion are the glass thermometers containing mercury or, in some cases, an alcohol. When temperature-indicating systems were first required for

* The name "degree Kelvin" ($°K$) was replaced by "kelvin" (K) in 1967.

use in aircraft it was obvious, from the nature of their construction and the method of observing their readings, that glass thermometers could not be employed particularly for temperature measurements in fluid systems for engines. As a result, indicators adopting the same liquid expansion principle but transmitting temperature information from a remote point were introduced. Similar transmitting-type thermometers utilizing the pressure of the vapour given off by a liquid at varying temperatures were also introduced. Some examples of both types of thermometer are still installed in certain small light aircraft, and it will not be out of place at this point to review briefly their construction and operation.

The liquid-expansion system consists of a steel bulb connected by a length of steel capillary tubing to a spiral Bourdon tube forming the measuring element. The system is completely filled with mercury. When the bulb is subjected to an increase in temperature the mercury expands and causes the Bourdon tube to unwind. The movement of the free end of the tube transmitted directly to a pointer.

With the exception of the measuring element, the components of a vapour-pressure thermometer are similar to those required for the liquid-expansion type. At normal ground atmospheric temperature, the bulb is partly filled with a volatile liquid, usually industrial ether. The remaining space contains only ether vapour. The capillary tube dips into the ether and is connected at its other end to a C-shaped Bourdon-tube measuring element. The capillary and Bourdon tubes are filled with liquid ether.

When the bulb is subjected to an increase of temperature, the ether vapour exerts a pressure on the liquid ether. As the bulb, capillary tube and Bourdon tube form a sealed system, the vapour pressure is transmitted by the liquid and causes the Bourdon tube to distend and actuate the pointer in the same manner as in a normal pressure gauge.

ELECTRICAL TEMPERATURE INDICATING SYSTEMS

As we have already noted, these systems fall into two main categories: variable-resistance and thermoelectric; the methods, are termed *resistance thermometry* and *pyrometry* respectively. In both cases, an understanding of the principles involved requires a knowledge of certain fundamental electrical laws and their applications.

Ohm's Law

This law may be stated as follows: *When a current flows in a conductor, the difference in potential between the ends of such conductor, divided by the current flowing through it, is a constant provided there is no change in the physical condition of the conductor.*

The constant is called the *resistance* (R) of the conductor and is measured in ohms (Ω). In symbols,

$$\frac{V}{I} = R \tag{1}$$

where V is the potential difference in volts (V), and I the current in amperes (A).

Calculations involving most conductors are easily solved by this law, for if any two of the three quantities (V, I and R) are known, the third can always be found. Thus, from eqn (1),

$$IR = V \tag{2}$$

$$\frac{V}{R} = I \tag{3}$$

A characteristic of the majority of metallic conductors is that their resistance increases when subjected to increases of temperature.

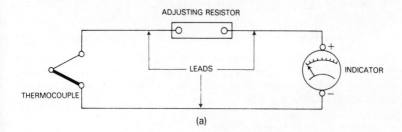

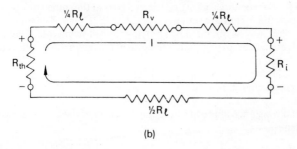

Fig 12.1 Series circuit
    (a) Practical circuit
    (b) Equivalent circuit
$$R_T = R_{th} + R_l + R_v + R_i$$

Resistances in Series

In all electrical instruments and associated circuits, we always find certain sections in which conductors are joined end to end. For example, in a thermoelectric temperature-indicating system, some of the essential conductors are joined as shown in Fig 12.1. A circuit formed in this manner is known as a *series circuit* and its combined, or total, resistance (R_T) is equal to the sum of the individual resistances. Thus

$$R_T = R_{th} + R_l + R_v + R_i$$

From a direct application of Ohm's law the current and the potential difference may also be derived:

$$I_T = \frac{V_T}{R_T}$$

this being the current at any point in the circuit. The total voltage (V_T) is equal to the sum of the voltage drops across the individual conductors and is always equal to the applied voltage, which in this case is the thermo-e.m.f. Thus

$$V_T = IR_{th} + IR_l + IR_v + IR_i = IR_T$$

Example. In a typical thermoelectric temperature-indicating circuit, the following resistance values apply: $R_{th} = 0.79\,\Omega$, $R_l = 24.87\,\Omega$, $R_v = 7\,\Omega$, and $R_i = 23\,\Omega$, so that

$$R_T = 0.79 + 24.87 + 7 + 23 = 55.66\,\Omega$$

The total voltage V_T is governed by the temperature at the thermocouple; therefore V_T and the total current I_T flowing through the indicator are variables. Assuming the temperature to be 500°C, the voltage generated by a typical thermocouple is 20·64 mV. Thus

$$I_T = \frac{V_T}{R_T} = \frac{20.64}{55.66} = 0.37\,\text{mA}$$

Resistances in Parallel

When two or more conductors are connected so that the same voltage is applied across each of them, they are said to form a *parallel circuit*. The total resistance of such a circuit may be obtained by applying the following rule. The reciprocal of the total resistance (R_T) is equal to the sum of the reciprocals of the individual conductor resistances.

Figure 12.2 shows a parallel circuit made up of six thermocouples, an arrangement which is employed for the measurement of turbine-engine exhaust gases.

If the total resistance of the thermocouple section of the circuit is represented by R_{TH}, then

$$\frac{1}{R_{TH}} = \frac{1}{R_{th1}} + \frac{1}{R_{th2}} + \ldots + \frac{1}{R_{th6}}$$

The resistances of the leads from the thermocouple junction box to the instrument terminals and the instrument resistance are represented by R_l and R_i respectively, and the circuit reduces to the simple series circuit shown. The total resistance R_T of the complete circuit can therefore be calculated from

$$R_T = R_{TH} + R_l + R_i$$

Assuming that all the thermo-e.m.f.s are equal, the voltage acting in the circuit will be equal to the thermo-e.m.f., V_{th}, of a single thermocouple, so that the current will be given by

$$I = \frac{V_{th}}{R_T}$$

Measurement of Temperature

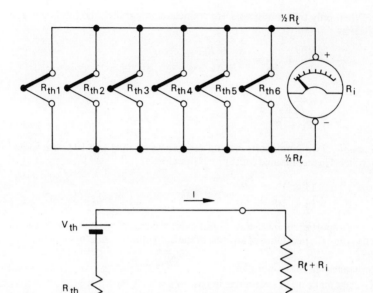

Fig 12.2 Parallel circuit

Example. The resistance of each of the thermocouples and its leads, shown in Fig 12.2, is 0·79 Ω; therefore, the total resistance R_{TH} of this part of the circuit is calculated as follows:

$$\frac{1}{R_{TH}} = \frac{1}{R_{th}} \times 6$$

so that

$$R_{TH} = \frac{R_{th}}{6} = \frac{0·79}{6} = 0·1317 \Omega$$

The leads resistance R_l is 24·87 Ω and the instrument resistance R_i is 23 Ω, so that the total circuit resistance is

$$R_T = R_{TH} + R_l + R_i$$
$$= 0·1317 + 24·87 + 23 = 48 \, \Omega \text{ very nearly}$$

Assuming that the thermocouples generate a voltage V_{th} of 20·64 mV (at 500°C) the total current will be

$$I = \frac{V_{th}}{R_T} = \frac{20·64}{48} = 0·43 \, \text{mA}$$

272 Aircraft Instruments

When only two conductors are connected in parallel, the total resistance is given by

$$\frac{1}{R_T} = \frac{1}{R_1} + \frac{1}{R_2} = \frac{R_1 + R_2}{R_1 R_2}$$

so that

$$R_T = \frac{R_1 R_2}{R_1 + R_2}$$

Example. Two conductors having resistances of 14 Ω and 10 Ω are joined in parallel. Their combined resistance is therefore

$$R_T = \frac{14 \times 10}{14 + 10} = \frac{140}{24} = 5 \cdot 833 \; \Omega$$

The important point to note about parallel circuits is that the total resistance is less than the resistance of any one of the individual conductors.

Resistances in Series-parallel

A circuit made up of resistances in series-parallel is shown in Fig 12.3 and is one based on an indicator used in conjunction with the thermocouple circuit of Fig 12.2. The solving of such a series-parallel circuit is done by the methods

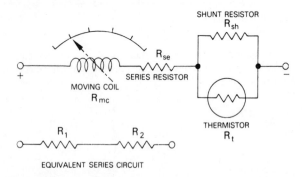

Fig 12.3 Series-parallel circuit

already illustrated and by (i) reducing each parallel group to an equivalent single resistance, (ii) solving the resulting series circuit for total resistance (R_T) and total current (I_T), (iii) obtaining the voltage drop across each parallel group, (iv) solving for all other quantities.

Example. The resistances of the four conductors in a typical indicator are $R_{mc} = 15 \cdot 5 \; \Omega$; $R_{se} = 2 \; \Omega$; $R_{sh} = 14 \; \Omega$ and $R_t = 10 \; \Omega$. R_{mc} and R_{se} form the series part of the circuit, thus

$$R_1 = R_{mc} + R_{se} = 15 \cdot 5 + 2 = 17 \cdot 5 \; \Omega$$

The parallel part of the circuit is formed by R_{sh} and R_t, thus

$$R_2 = \frac{R_{sh}R_t}{R_{sh} + R_t} = \frac{14 \times 10}{14 + 10} = \frac{140}{24} = 5 \cdot 833 \, \Omega$$

Reducing the circuit to an equivalent series circuit, the total resistance R_T is

$$R_T = R_1 + R_2 = 17 \cdot 5 + 5 \cdot 833 = 23 \cdot 333 \, \Omega$$

Factors Governing Resistance

The resistance of a single conductor is governed by three main factors: (i) the material of which it is composed, (ii) its dimensions (length and cross-sectional area), and (iii) its physical state, particularly its temperature.

The *resistivity* ρ of a conductor is the resistance of a sample of the material having unit length and unit cross-sectional area. At a given temperature, it is therefore a constant for a given material and is usually expressed in microhms per centimetre cube.

At a given temperature, the resistance R of a material is directly proportional to the length l in centimetres, and inversely proportional to the cross-sectional area a in square centimetres. Thus

$$R = \rho \frac{l}{a}$$

Resistance and Temperature

The resistance of a conductor is dependent on temperature, the effect of which is to change the dimensions by thermal expansion and to change the resistivity. The first effect is comparatively small, the main changes of resistance being due to changes of ρ. In the case of pure metallic conductors, resistance increases with increase in temperature, and this is the basis of temperature measurement in resistance thermometry.

The two metals most commonly used in aircraft resistance thermometry are nickel and platinum, both of which are manufactured to a high degree of purity and reproducibility of resistance characteristics.

The approximate resistance of a metal at a temperature t is given by the linear equation

$$R_t = R_0(1 + \alpha t) \tag{1}$$

where R_t is the resistance at temperature $t\,°C$; R_0 the resistance at $0°C$, and α is the *temperature coefficient of resistance*.

In evolving temperature/resistance calibration laws, however, the above simple linear relation cannot be applied because the temperature coefficients of nickel and platinum can only be regarded as constant over the temperature range 0–100°C. Therefore, in order to obtain greater accuracy and an extension of the temperature range to be measured, it is necessary to introduce another constant (β) into eqn (1) so that the resistance is represented by the quadratic equation

$$R_t = R_0(1 + \alpha t + \beta t^2) \tag{2}$$

The number of constants involved depends on the temperature range, but in general, two are sufficient up to about 600°C.

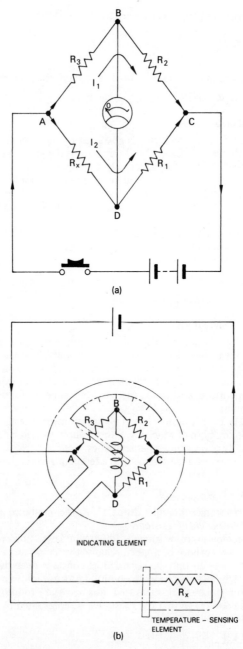

Fig 12.4 Wheatstone bridge network
     (a) Basic circuit   (b) Application to temperature measurement

WHEATSTONE BRIDGE

The most common method of measuring resistance is by means of the well-known *Wheatstone bridge* network shown in Fig 12.4 (*a*).

The circuit is made up of four resistances arms R_1, R_2, R_3 and R_x. A moving-coil or moving-spot galvanometer is connected across points B and D, and a source of low voltage is connected across points A and C. Current flows in the directions indicated by the arrows, dividing at point A and flowing through R_3 and R_x at strengths which we may designate respectively as I_1 and I_2. At point C the currents reunite and flow back to the voltage source.

Let us assume that the resistances of the four arms of the bridge are so adjusted that B and D are at the same potential; then no current will flow through the galvanometer and so it will read zero. Under these conditions the bridge is said to be "balanced." This can be shown by applying Ohm's law; if the potential difference between A and B or A and D is, say, V_1, and the potential difference between B and C or D and C is V_2, then

$$V_1 = I_1 R_3 = I_2 R_x \tag{3}$$

and

$$V_2 = I_1 R_2 = I_2 R_1 \tag{4}$$

If now eqn (3) is divided by eqn (4), we have

$$\frac{V_1}{V_2} = \frac{R_3}{R_2} = \frac{R_x}{R_1} \tag{5}$$

from which

$$R_x = \frac{R_1 R_3}{R_2} \tag{6}$$

Hence, an unknown resistance can be calculated by adjusting the values of the three others until no current flows through the galvanometer, as indicated by no movement of its pointer or spot.

It will be apparent that, if the resistor R_x is subjected to varying temperatures and its corresponding resistances are determined, then it is feasible for the network to serve as a simple electrical-resistance thermometer system.

Obviously some rearrangement of the circuit is necessary in order to obtain automatic response to the variations in the temperature/resistance relation; the manner in which this can be effected is illustrated in Fig 12.4 (*b*). The unknown resistance R_x forms the temperature-sensing element and is contained within a metal protective sheath, the assembly being called a bulb. The three other resistances instead of being adjustable, are fixed and are contained within the case of a moving-coil indicating element calibrated in units of temperature. Both components are suitably interconnected and supplied with direct current.

When the bulb is subjected to temperature variations, there will be a corresponding variation in R_x. This upsets the balance of the indicator circuit, and the value of R_x at any particular temperature will govern the amount of current flowing through the moving coil. Thus, for a given value of R_x, the out-of-balance current is a measure of the prevailing temperature.

It will be noted that there is an important difference between the two applications of this bridge network. In one the measurement of a resistance is dependent upon the circuit current being in balance, while in the other it is measured in terms of out-of-balance current.

There is in fact only one point in a bridge type of temperature indicator at which the circuit is balanced and at which no current will flow through the moving coil; this is known as the *null point*. It is usually indicated on the scale of the instrument by means of a small datum mark and the pointer takes up this position when the power supply is disconnected.

For accurate temperature measurement, however, this form of indicating circuit has the disadvantage that the out-of-balance current also depends on the voltage of the power supply. Hence, errors in indicated readings will occur if the voltage differs from that for which the instrument was initially calibrated. Bridge-type thermometers have therefore been largely superseded by those operating on what is called the ratiometer principle described on page 280.

TEMPERATURE-SENSING ELEMENTS

The general arrangement of a temperature-sensing element commonly used for the measurement of liquid temperatures is shown in Fig 12.5. The resistance coil is wound on an insulated former and the ends of the coil are connected to a two-pin socket via contact strips. The bulb, which serves to protect and seal the

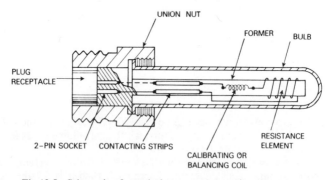

Fig 12.5   Schematic of a typical temperature-sensing element

element, may either be a brass or stainless-steel tube closed at one end and soldered to a union nut at the other. The union nut is used for securing the complete element in the pipeline or component of the system whose liquid temperature is required. The two-pin socket is made a tight fit inside the male portion of the union nut, the receptacle of which ensures correct location of the socket's mating plug.

It will be noted from the diagram that the coil is wound at the bottom end of its former and not along the full length. This ensures that the coil is well immersed in the hottest part of the liquid, thus minimizing errors due to radiation and conduction losses in the bulb.

A calibrating or balancing coil is normally provided so that a standard constant temperature/resistance characteristic can be obtained, thus permitting interchangeability of sensing elements. In addition the coil compensates for any slight change in the physical characteristics of the element. The coil, which may be made from Manganin or Eureka, is connected in series with the sensing element and is adjusted by the manufacturer during initial calibration.

ELECTRICAL TEMPERATURE INDICATORS

The measuring elements employed in resistance-type and thermoelectric temperature-indicating systems depend for their operation on the fact that electric current flowing through a conductor produces a magnetic field in and around the conductor.

In order to utilize this effect as a method of measurement, it is necessary to have a free-moving conductor in a magnetic field which is both permanent and of uniform strength. In this manner, as will be shown, advantage can be taken of the interaction between the two magnetic fields and the resultant forces to move the conductor and its indicating element to definite positions.

Conductor Carrying Current in a Magnetic Field

When a current-carrying conductor is placed in a magnetic field, the interaction of the field produced by the current and the field in which the conductor is located exerts a force upon the conductor. This force is in direct proportion to the flux density, the current and the length of the conductor.

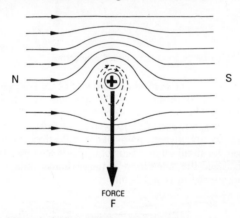

Fig 12.6 Magnetic field around a straight conductor

In Fig 12.6 the field set up around the conductor is shown as being in the same direction as the main field above the conductor and in the reverse direction to the main field below the conductor. Thus, the interaction is similar to that of like and unlike poles of two bar magnets, i.e. lines of magnetic force flowing in the same direction attract each other while those flowing in opposite directions repel each other.

The lines of force of the main field thus become distorted so that it is stronger on one side of the conductor than on the other, the net effect being that the conductor is subjected to a force in the direction of the weaker field. If the conductor is free to move in the main field, as is the case with electric motors and moving-coil instruments, then it is clear that the motion will be in the same direction as the force. The force is also dependent upon the angle which the conductor makes with the main magnetic field, being maximum when it is at right angles to the field.

The Moving-coil Indicator

Consideration thus far has been given to a straight current-carrying conductor fixed in a permanent magnetic field. If the conductor is formed into a single coil and pivoted at a point P as shown in Fig 12.7, then, for the direction of current

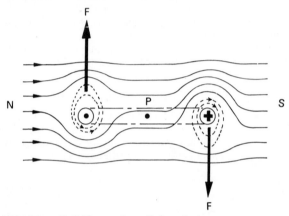

Fig 12.7 Magnetic field around a coiled conductor

indicated, forces, F,F will be exerted on each side of the coil, each producing a torque Fr, so that the total torque will be $2Fr$ and will cause the coil to rotate in a clockwise direction. This is the basis of any moving-coil indicator.

In the practical case, however, it is necessary to intensify the forces acting on the coil in order to obtain reasonably large deflections of the coil and its indicating element for small values of current. This is accomplished by placing a soft-iron core between the pole pieces of the permanent magnet, and also by increasing the number of turns of the coil.

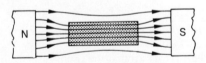

Fig 12.8 Effect of soft-iron in a magnetic field

Soft iron has a lower reluctance to lines of force than air; therefore, when it is placed in a uniform magnetic field as indicated in Fig 12.8, the lines of force from its surroundings are drawn into the iron and it becomes magnetized by induction.

Referring to Fig 12.9 we see that by shaping the pole pieces and soft-iron core cylindrically, a radial magnetic field of even greater intensity and uniformity is obtained in the narrow air gap in which the sides of the coil rotate.

Measurement of Temperature

The constructional details of a typical moving-coil indicator are shown in Fig 12.10. The permanent magnet is made from a special alloy possessing high remanence and coercive force characteristics, and may be either circular or rectangular in shape, and machined to size within very close tolerances. An adjustable shunt is secured across the pole pieces to vary the field in the air gap during calibration of the indicator.

The cylindrical soft-iron core and the moving coil assembly are usually built up into a single unit which can be positioned in the magnet air gap. In the example shown, the unit is secured by screwing a bridge piece to the magnet pole pieces.

The coil consists of a number of turns of fine copper wire wound on a rectangular aluminium former or frame, pivoted in jewelled bearings so that when current flows through the coil the combination of magnetic fields set up around each turn will increase the force required to deflect the coil.

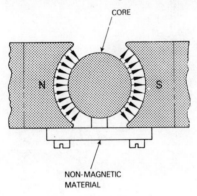

Fig 12.9 Effect of a cylindrical soft-iron core

Current is supplied to the moving coil via two flat-coil hairsprings the main function of which is to ensure that currents of varying magnitudes shall produce proportionate deflections of the coil. In other words, they form the *controlling system*, an essential part of any moving-coil instrument without which the coil would move to its maximum deflected position regardless of the magnitude of the current, and moreover would not return to its zero position on cessation of the current flow. The hairsprings are made of materials having low resistance and low temperature coefficients, phosphor-bronze and beryllium-bronze being most commonly used. The effects of extreme temperature variations are minimized by coiling the springs in opposite directions so that they act one against the other.

The setting of the moving coil and pointer to zero is carried out by means of an adjusting device similar to that described in Chapter 2 (page 18).

As in all indicating instruments, mechanical balance of the moving system is necessary to ensure uniform and symmetrical wear on pivots and bearings, so preventing out-of-balance forces from causing errors in indication. This is usually effected by attaching balance arms on the supporting spindle and providing them with either adjustable balance weights or wire coiled round and soldered to the arms.

A further essential requirement of any moving-coil instrument is that the coil should take up its deflected position without oscillation and overshoot, i.e. it must be damped and rendered *dead beat* in its indications. In this type of instrument damping is effected automatically by the aluminium former on which the coil is wound. As the former moves in the air gap it cuts the magnetic flux, thus setting up eddy currents within itself to produce a force opposing that causing movement of the coil. Since the force is proportional to the velocity of the moving coil, the latter will be retarded so as to take up its deflected position without overshoot.

The complete instrument movement is enclosed in a soft-iron case to screen it against the effects of external magnetic fields.

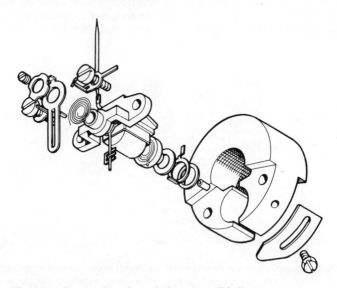

Fig 12.10   Construction of a typical moving-coil indicator

Ratiometer System

A ratiometer-type temperature-indicating system consists of a sensing element and a moving-coil indicator, which unlike the conventional type has two coils moving together in a permanent-magnet field of non-uniform strength. The coil arrangements and methods of obtaining the non-uniform field depends on the manufacturer's design, but three methods at present in use are shown in Fig 12.11.

Figure 12.12 shows the circuit in basic form, and from this it will be noted that two parallel resistance arms are formed; one containing a coil and a fixed calibrating resistance R_1, and the other containing a coil in series with a calibrating resistance R_2 and the temperature-sensing element R_x. Both arms are supplied with direct current from the aircraft's main power source, but the coils are so wound that current flows through them in opposite directions (see also Fig 12.11). As in any moving-coil indicator, rotation of the measuring element is produced by forces which are proportional to the product of the current and field strength, and the direction of rotation depends on the direction of current relative to the magnetic field. In a ratiometer, therefore, it follows that the force produced by one coil will always tend to rotate the measuring element in the opposite direction to the force produced by the second coil, and furthermore, as the magnetic field is of non-uniform strength, the coil carrying the greater current will always move towards the area of the weaker field, and vice versa.

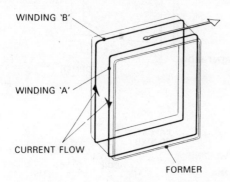

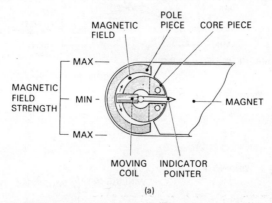

Fig 12.11   Methods of obtaining a non-uniform magnetic field
(a) Parallel coil   (continued overleaf)

For purposes of explanation, let us assume that the basic circuit of Fig 12.12 employs an instrument which utilizes the crossed-coil winding method shown in Fig 12.11(b), and that winding B is in the variable-resistance arm, and winding A is in the fixed-resistance arm. The resistances of the arms are so chosen that at the zero position of the instrument the forces produced by the currents flowing in each winding are in balance. Although the currents are unequal at this point, and indeed at all other points except mid-scale, the balancing of the torques is always produced by the strength of the field in which the windings are positioned.

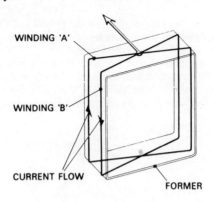

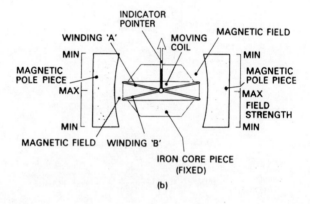

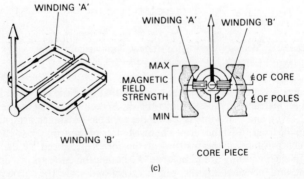

Fig 12.11 (contd.)
    (b) Cross coil   (c) Twin former

When the temperature at the sensing element R_x increases, then in accordance with the temperature/resistance relationship of the material used for the element, its resistance will increase and so cause a decrease in the current flowing in winding B and a corresponding decrease in the force created by it. The current ratio is therefore altered and the force in winding A will rotate the measuring element so that both windings are carried round the airgap; winding B being advanced further into the stronger part of the magnetic field while winding A is being advanced to a weaker part. When the temperature at the sensing element stabilizes at its new value the forces produced by both windings will once again balance, at a new current ratio, and the angular deflection of the measuring element will be proportional to the temperature change.

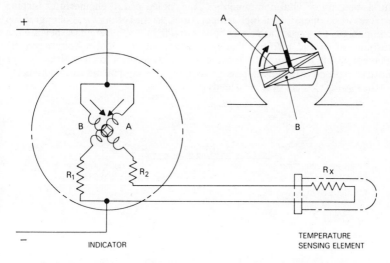

Fig 12.12 Basic ratiometer circuit

When the measuring element is at the mid-position of its rotation, the currents in both windings are equal since this is the only position where the two windings can be in the same field strength simultaneously.

In a conventional moving-coil indicator, the controlling system is made up of hairsprings which exert a controlling torque proportional to the current flowing through the coil. Therefore, if the current decreases due to a change in the power supply applied to the indicator, the deflecting torque will be less than the controlling torque of the springs and so the coil will move back to a position at which equilibrium between torques is again established. The pointer will thus indicate a lower reading.

A ratiometer system, however, does not require hairsprings for exerting a controlling torque, this being provided solely by the appropriate coil winding and non-uniform field arrangements. Should variations in the power supply occur they will affect both coils equally so that the ratio of currents flowing in the coils remains the same and tendencies for them to move to positions of differing field strength are counterbalanced.

In practical applications of the ratiometer system, a spring is, in fact, used and so at first sight this may appear to defeat the whole object of the ratiometer principle. It is, however, essential that the moving-coil former and pointer should take up an off-scale position when the power supply is disconnected, and this is the sole function of the spring. Since it exerts a very much lower torque than a conventional moving-coil indicator control spring, its effects on the ratiometer controlling system and indication accuracy, under power supply changes normally encountered, are very slight.

THERMOELECTRIC THERMOMETERS

Thermoelectric thermometers play an important part in monitoring the structural integrity of vital components of air-cooled piston engines and turbine engines when operating at high temperatures. In the former class of engine the components concerned are the cylinders, while in turbine engines they are the turbine rotors and blading. The systems consist of a thermocouple sensing element which, depending on the application, is secured to an engine cylinder head or exposed to turbine exhaust gases, and a moving-coil indicator connected to the sensing element by special leads.

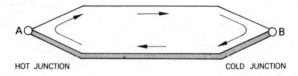

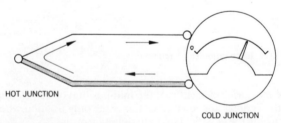

Fig 12.13 Thermocouple principle

Thermocouple Principle

Theremoelectric temperature-measuring instruments depend for their operation on electrical energy which is produced by the direct conversion of heat energy at the measuring source. Thus, unlike resistance thermometers, they are independent of any external electrical supply.

This form of energy conversion, known as the *Seebeck effect*, was first demonstrated by Seebeck in 1871, when he discovered that by taking two wires of dissimilar metals and joining them at their ends, so as to form two separate junctions, A and B as in Fig 12.13, a thermo-e.m.f. was produced when the junctions were maintained at different temperatures, causing current to flow round the circuit.

The arrangement of two dissimilar metal wires joined together in this manner is called a *thermocouple*; the junction at the higher temperature being conventionally termed the hot or measuring junction, and that at the lower temperature the cold or reference junction. (In practice, the hot junction is in the form of a separate unit for sensing the temperature and this is regarded generally as the thermocouple proper.)

Experiments which followed Seebeck's discovery proved the existence of two other effects.

When an electric current flows through the junction of two different substances it causes heat to be either absorbed or liberated at the junction, depending on the direction of the current. This is known as the *Peltier effect*. In a circuit in which the only generated voltage is a thermo-e.m.f., current flows through a heated junction in a certain direction. If, instead of heat being supplied to the system, a battery is introduced into the circuit, of such polarity that a current is driven in the same direction as the thermo-current, the junction which was previously heated will be cooled, and the junction which was previously held at a constant temperature will be heated.

Lord Kelvin (when he was Sir William Thomson) discovered that effects similar to the Seebeck and Peltier effects occur in a single, homogeneous conductor: if two parts of a conductor are at different temperatures, an e.m.f. is generated; and when current flows from a part of a conductor to another which is at a different temperature, heat may be either liberated or absorbed. These phenomena are different aspects of a single, *Thomson effect*.

Thermocouple Materials and Combinations

The materials selected for use as thermoelectric sensing elements fall into two main groups, *base metal* and *rare metal*, and are listed in Table 12.1. The choice of a particular thermocouple is dictated by the maximum temperature to be encountered in service. Thermocouples required for use in aircraft are confined to those of the base-metal group.

In order to utilize the thermoelectric principle for temperature measurement, it is obviously necessary to measure the e.m.fs. generated at the various temperatures. This is done by connecting a moving-coil millivoltmeter, calibrated in degrees Celsius, in series with the circuit so that it forms the cold junction. The introduction of the instrument into the circuit involves the presence of additional junctions which produce their own e.m.f.s. and so introduce errors in measurement. However, the effects are taken into consideration when designing practical thermocouple circuits, and any errors resulting from "parasitic e.m.fs.," as they are called, are eliminated.

The dial presentations of typical indicators are illustrated in Fig 12.14. The SET marking on the exhaust-gas temperature indicator dial indicates the temperature at which the pointer is positioned during the setting-up procedure carried out after installation.

Types of Thermocouple

The thermocouples employed in aircraft thermoelectric indicating systems are of two basic types (i) *surface contact* and (ii) *immersion*. Typical examples are shown in Figs 12.15(*a*) and (*b*).

Table 12.1 Thermocouple Combinations

Group	Metals and composition		Maximum temperature °C (continuous)	Application
	Positive wire	Negative wire		
Base metal	Copper (Cu)	Constantan (Ni, 40%; Cu, 60%)	400	Cylinder-head temperature measurement
Base metal	Iron (Fe)	Constantan (Ni, 40%; Cu, 60%)	850	
Base metal	Chromel (Ni, 90%; Cr. 10%)	Alumel (Ni, 90%; Al, 2% + Si + Mn)	1,100	Exhaust-gas temperature measurement
Rare metal	Platinum (Pt)	Rhodium-platinum (Rh, 13%; Pt, 87%)	1,400	Not utilized in aircraft temperature-indicating systems

Cr. chromium; Ni, nickel: Al, aluminium; Si, silicon; Mn, manganese

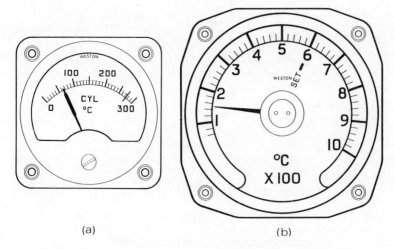

Fig 12.14 Dial presentations of thermoelectric temperature indicators
(a) Engine-cylinder-head temperature indicator
(b) Turbine-engine gas temperature indicator

The surface-contact type is designed to measure the temperature of a solid component and is used as the temperature-sensing element of air-cooled-engine cylinder-head temperature-indicating systems. The copper/constantan or iron/constantan element may be in the form of a "shoe" bolted in good thermal contact with a cylinder head representative of the highest temperature condition or in the form of a washer bolted between the cylinder head and a sparking plug.

The immersion type of thermocouple is designed for the measurement of gases and is therefore used as the sensing element of turbine-engine gas temperature-indicating systems. The chromel/alumel hot junction and wires are encased in ceramic insulation within a metal protection sheath, the complete assembly forming a probe which can be immersed in the gas stream at the points selected for measurement.

Immersion-type thermocouples are further classified as *stagnation* and as *rapid response*, their application depending upon the velocity of the engine exhaust gases. In pure jet engines the gas velocities are high, and so in these engines stagnation thermocouples are employed. The reason for this will be clear from Fig 12.15(c), which shows that the gas entry and exit holes, usually called sampling holes, are staggered and of unequal size, thus slowing up the gases and causing them to stagnate at the hot junction, thus giving it time to respond to changes of gas temperature.

Rapid-response thermocouples are employed in turboprop engines since their exhaust-gas velocities are lower than those of pure jet engines. As can be seen from Fig 12.15(d), the sampling holes are diametrically opposite each other and of equal size; therefore the gases can flow directly over the hot junction enabling it to respond more rapidly. Typical response times for stagnation and rapid-response thermocouples are 1 to 2 seconds and 0·5 to 1 second, respectively.

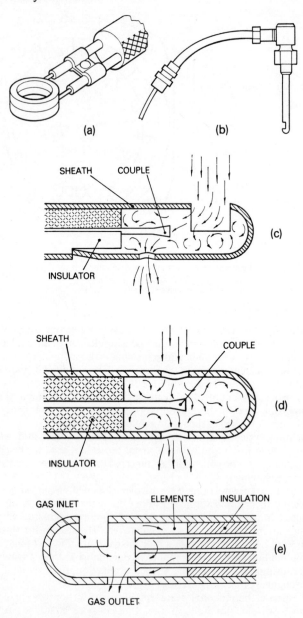

Fig 12.15 Types of thermocouple
(a) Surface contact
(b) Immersion
(c) Stagnation
(d) Rapid response
(e) Triple-element probe

Thermocouple probes are also designed to contain double and/or triple thermocouple elements as shown in Fig 12.15(e). The purpose of such arrangements is to provide additional temperature signals for engine installations utilizing an exhaust-gas temperature-control system (see page 346) and an engine combustion-analyzer system.

When the hot junctions of immersion-type thermocouples are in contact with the gas stream, it is obvious not only that the stream velocity will be reduced but also that the gas will be compressed by the expenditure of kinetic energy, resulting in an increase of hot-junction temperature. It is in this connection that the term *recovery factor* is used, defining the proportion of kinetic energy of the gas recovered when it makes contact with the hot junction. This factor is, of course, taken into account in the design of thermocouples so that the "heat transfer," as we may call it, makes the final reading as nearly as possible a true indication of total gas temperature.

In Fig 12.16 the constructional details of a third type of immersion thermocouple are shown; this has been designed to measure the gas temperatures between turbine stages. The hot junction is housed inside a sheath which is shaped so as to form the leading edge of a stator guide vane, and for this reason the assembly is usually called a *nozzle-guide-vane* thermocouple.

Gases flow over the hot junction which is positioned between sampling holes of equal diameter as in a rapid-response thermocouple. Unlike the latter, however, nozzle-guide-vane thermocouples do not exhibit the same characteristics, because the sampling holes are much smaller in diameter, and furthermore the couple response is made slower by the mass of the sheath and its proximity to the guide vane.

Location of Exhaust Gas Thermocouple Probes

The points at which the gas temperature of an engine is to be measured are of great importance, since they will determine the accuracy with which measured temperatures can be related to engine performance. The ideal location for a thermocouple hot junction would be either at the turbine blades themselves or at the turbine entry, but certain practical difficulties are involved which preclude such locations. Consequently, the locations generally selected are at the exhaust, or jet pipe unit, and between the turbine stages at one of the stator positions. The temperatures at these locations are much lower, but they relate very closely to those at the turbine entry.

For accurate measurement it is necessary to sample temperatures from a number of points evenly distributed over a cross-section of the gas flow. This is because temperature differences can exist in various zones or layers of the flow through the turbine and exhaust unit, and so measurement at one point only would not be truly representative of the conditions prevailing.

Therefore, the measuring system always consists of a group of six or more thermocouple probes, depending on the engine type, suitably disposed in the gas flow, and connected in parallel so as to measure a good average temperature condition (see Fig 12.17). The thermocouples and their leads are built up into a harness assembly designed for each particular type of engine. Nozzle guide-vane thermocouples are arranged in pairs of long-reach and short-reach probes, named according to the extent to which the hot junctions and gas sampling holes reach into the gas stream.

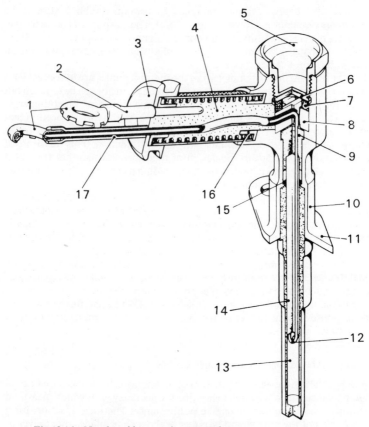

Fig 12.16 Nozzle guide vane thermocouple

1	Terminals	10	Body
2	Nickel-aluminium lead	11	Mounting flange
3	End-cap	12	Couple junction
4	Silicone filler	13	Probe
5	Plug	14	Ceramic insulator
6	Mica washers	15	Washer
7	Ceramic top clamp	16	Ferrule
8	Ceramic bottom clamp	17	Nickel-chromium lead
9	Glass seal		

Cold Junction Temperature Compensation

As we have already seen, the indicator of a thermoelectric temperature measuring system forms the cold junction of the system, and the e.m.f. produced depends upon the difference between the temperature of this junction and the hot junction. It is thus apparent that, if the ambient temperature of the indicator itself changes while the hot junction temperature remains constant, then a change in e.m.f. will result causing the indicator to read a different temperature.

In applying this principle to the measurement of aircraft engine temperatures, such temperature differences constitute indication errors which cannot be tolerated, since it is essential for the indicated readings to be representative of the temperature conditions at the hot junction only. In order to achieve such readings it is necessary to provide indicators with a device which will detect ambient temperature changes and compensate for possible errors; such a device is called a *cold-junction compensator*. Before going into the mechanical and operating details of a compensator, it is useful to consider first how the changes in e.m.f. actually arise.

The various combinations of thermocouple materials specified for use in aircraft, conform to standard temperature/e.m.f. relationships, and the indicators employed in conjunction with these combinations are calibrated accordingly. The e.m.f.s obtained correspond to a cold-junction temperature which is usually maintained at either 0°C or 20°C.

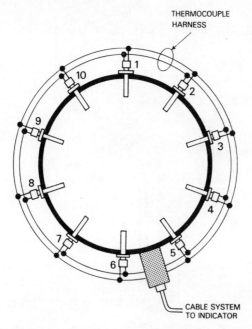

Fig 12.17   Disposition of exhaust-gas thermocouples

Let us assume, for example, that the cold junction is maintained at 0°C and that the hot junction temperature has reached 500°C. At this temperature difference a standard value of e.m.f. generated by a chromel/alumel combination is 20·64 mV. If now the temperature at the cold junction increases to 20°C while the hot junction remains at 500°C, the temperature difference decreases to 480°C and the e.m.f. equivalent to this difference is now 20·64 mV minus the e.m.f. at 20°C, and as a standard value this corresponds to 0·79 mV. Thus, the moving element of the indicator will respond to an e.m.f. of 19·85 mV and move "down scale" to a reading of 480°C.

A change, therefore, in ambient and cold junction temperature decreases or increases the e.m.f. generated by the thermocouple, making the indicator read low or high by an amount equal to the change of ambient temperature.

The method commonly adopted for the compensation of these effects is quite simple and is, in fact, an adaptation of the bimetallic strip principle described on page 19. For this purpose, however, the strips of dissimilar metals are fastened together and coiled in the shape of a flat spiral spring. One end of the spring is anchored to a bracket which forms part of the moving-element support, while the other end (free end) is connected by an anchor tag to the outer end of one of the controlling hairsprings, thus forming the fixed point for the hairspring. A typical arrangement is shown in Fig 12.18.

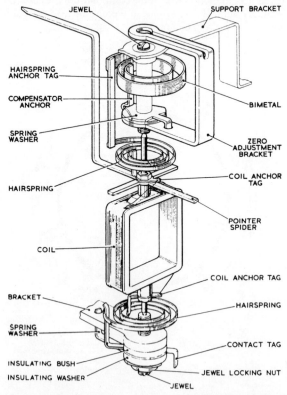

Fig 12.18 Application of bimetal type of cold junction compensator

When the indicator is on open-circuit, i.e. disconnected from the thermocouple system, the spring responds to ambient temperature changes at the indicator, an increase in temperature causing the spring to unwind so that its free end carries the hairspring and moving element round to indicate the increase in temperature. Conversely, a temperature decrease will wind up the compensator spring so that the moving element will indicate the lower temperature. Therefore an indicator disconnected from its e.m.f. source operates as a direct-reading bimetallic type of thermometer.

With the thermocouple system connected to the indicator the circuit is completed, and if the two junctions are at the temperatures earlier assumed, namely

0°C and 500°C, the e.m.f. will position the moving element to read 500°C. If the temperature at the indicator increases to 20°C, then, as already illustrated, the e.m.f. is reduced but the tendency for the moving element to move down scale is now directly opposed by the compensating spring as it unwinds in response to the 20°C temperature change. The indicator reading therefore remains at 500°C, the true hot junction temperature.

Compensation of Moving-coil Resistance Changes

The changes of thermocouple e.m.f. so far discussed are not, unfortunately, the only effect brought about by changes of ambient temperature at the indicator. A second effect requiring compensation is the change in resistance of the moving coil itself. The manner in which this is accomplished depends upon the particular design, but the two methods most commonly adopted are the thermo-magnetic shunt and thermo-resistor described in Chapter 2.

External Circuit and Resistance

The external circuit of a thermoelectric indicating system consists of the thermocouple and its leads, and the leads from the junction box at the engine bulkhead to the indicator terminals. From this point of view, it might therefore be considered as a simple and straightforward electrical instrument system. However, whereas the latter may be connected up by means of the appropriate copper leads or cable normally used in aircraft, it is not acceptable to do so in a thermoelectric system.

This may be explained by taking the case of a copper/constantan thermocouple which is to be connected to a cylinder-head temperature indicator. If a length of normal copper twin-core cable is connected to the thermocouple terminal box, then one copper lead will be joined to its thermocouple partner, but the other one will be joined to the constantan lead. It is thus apparent that the joining of the two dissimilar metals introduces another effective hot junction which will respond to temperature changes occurring at the junction box, and in unbalancing the temperature/e.m.f. relationship will cause the indicator readings to be in error. Similarly, all terminal connections which may be necessary for routing the leads through the aircraft and connections at the indicator itself will create additional hot junctions and so aggravate indicator errors.

In order to eliminate these hot junctions, it is the practice to use leads made of the same materials as the thermocouple itself; such leads being known as *extension leads*. It is sometimes the practice also to use thermocouple materials having similar thermoelectric characteristics in combination; for example, a chromel/alumel thermocouple may be joined to its indicator by copper/constantan leads known as *compensating leads*. These not only compensate for parasitic effects but also reduce the cost, since chromel-alumel is expensive.

Additional hot junctions at the indicator itself are eliminated by making the positive terminal of copper or brass, and the negative terminal of constantan. This applies to indicators for use with the various thermocouple combinations.

Another important factor in connection with the external circuit is its resistance, which must be kept not only low but also constant for a particular installation. Indicators are normally calibrated for use with either an $8\,\Omega$ or a $25\,\Omega$ external circuit resistance and are marked accordingly on their dials. The thermocouple, leads and harness are made up in fixed low-resistance lengths, and for turbine engines they form component parts of the engine. Similarly, the extension or compensating leads must also be made up in lengths and of uniform resistance to suit the varying distances between thermocouple hot junction and indicator installations.

Adjustment of the total external circuit resistance to the value for which associated instruments have been calibrated is made possible by connecting a trimming resistance spool in series with one of the extension or compensating leads. The material used for the spool may be either Eureka or Manganin, both of which have negative temperature coefficients of resistance. The inclusion of the spool in the circuit introduces another junction, of course, and whether it should be in the positive lead or the negative lead is governed by the material of the spool and the leads. For example, if a Eureka spool is to be used in a copper/constantan thermocouple system, it must be connected in the constantan or negative lead, whereas a Manganin spool must be connected in the copper or positive lead, in order to have negligible thermoelectric effect. In a chromel/alumel system, Manganin is usually employed, and for this circuit negligible thermoelectric effect is obtained by connecting it in the chromel or positive lead.

It is possible for the resistance of the extension leads to change with changes of temperature, but as most of the total circuit resistance is within the indicator, the compensation methods described earlier can also be effective in compensating for changes in external circuit resistance.

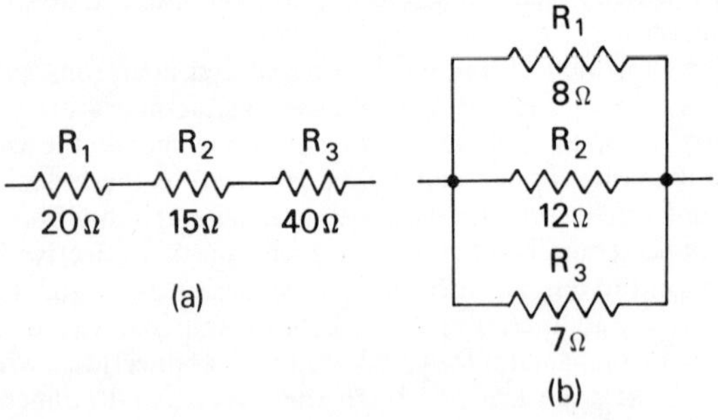

Fig 12.19

QUESTIONS

12.1 Describe how temperature can cause variations in the properties of substances.

12.2 Define the following: (i) convection, (ii) conduction, (iii) radiation, (iv) temperature.

12.3 Define the fundamental interval of temperature and state how it is divided in accordance with the Celsius and Fahrenheit scales.

12.4 (a) What is the relationship between the Celsius, Fahrenheit and absolute [Kelvin] temperature scales?
(b) Convert 40°C into degrees absolute and degrees Fahrenheit. (SLAET)

12.5 Convert 77°F into degrees Celsius.

12.6 Define Ohm's law and include the expressions used for calculating the three quantities, voltage, current and resistance.

12.7 Calculate the resistance (R_T) of circuits (a) and (b) in Fig 12.19. What current will flow in each circuit when 24 volts d.c. is applied?

12.8 (a) Calculate the resistance (R_T) of a circuit containing two resistances in parallel, the values being $R_1 = 20\ \Omega$ and $R_2 = 45\ \Omega$.
(b) Draw a diagram to illustrate a series-parallel circuit and explain how you would solve such a circuit from given values.

12.9 (a) What are the factors governing the resistance of a conductor?
(b) A round copper conductor 0·8mm diameter has a length of 1,200 cm. Calculate its resistance at normal temperature. (The resistivity of copper is approximately $1·7\mu\Omega$-cm.)

12.10 Draw a diagram to illustrate a Wheatstone bridge circuit, and from this derive the expression for calculating the value of an unknown resistance.

12.11 (a) Explain how the Wheatstone bridge circuit may be utilized for the measurement of temperatures.
(b) Would the circuit be in balance at each temperature?

12.12 (a) What materials are most commonly used for resistance-type temperature-sensing elements?
(b) Describe the construction of a typical element.

12.13 Sketch and describe the construction of a moving-coil meter. (SLAET)

12.14 (a) Explain the operating principle of a moving-coil meter.
(b) Does such an instrument have a linear or a non-linear scale?

12.15 Why is a soft-iron core placed within the coil of a moving-coil instrument?
(SLAET)

12.16 Explain how instruments are screened against the effects of external magnetic fields.

12.17 (a) Describe the construction and operation of a ratiometer type of temperature indicator.
(b) What principal advantage does the instrument have over a Wheatstone bridge type?

12.18 What would be the effect of an open-circuit between the transmitter and indicator of a ratiometer system? (SLAET)

12.19 (a) Explain the thermocouple principle, and state to what temperature measurement it is applied.
(b) What metal combinations are used in aircraft pyrometry systems?

12.20 A thermocouple used for the measurement of engine cylinder-head temperature measurement is of: (a) the immersion type, (b) the stagnation type, (c) the surface contact type. Which of these statements is correct?

12.21 (a) Describe the construction of a typical thermocouple assembly used for turbine-engine exhaust-gas temperature measurement.
(b) How is it ensured that a good average temperature condition of exhaust gases is measured?

12.22 What effects can changes of cold-junction temperature have on the indications of thermoelectric indicators? Describe the methods of compensation.

12.23 What is the difference between extension leads and compensating leads?

12.24 (a) What is the purpose of the resistor connected in series with one of the leads in certain pyrometry systems?
(b) Name the materials from which resistors are made and explain why they are connected in different leads.

13 Measurement of Pressure

In many of the systems associated with the operation of aircraft and engines, liquids and gases are used the pressures of which must be measured and indicated. The gauges and indicating systems fall into two main categories: (i) *direct-reading*, or those to which the source of pressure is directly connected, and (ii) *remote-indicating*, or those connected to a pressure source at some remote point, the pressure being transmitted to the gauge via a special fluid or an electric current.

Before going into the details of construction and components required, we shall first consider the methods by which pressure can be measured.

METHODS OF MEASURING PRESSURE

Pressure, which is defined as force per unit area, may be measured directly either by balancing it against that produced by a column of liquid of known density, or it may be permitted to act over a known area and then measured in terms of the force produced. The former method is the one utilized in simple *U-tube manometers*, while the second enables us to measure the force by balancing it against a known weight, or by the strain it produces in an elastic material.

U-TUBE MANOMETER

The simple U-tube manometer shown in Fig 13.1, consists of a glass tube partially filled with a liquid, usually water or mercury, which finds its own level at a point 0 within the open-ended limbs of the U. If a low-pressure source is connected to the limb A, then a force equal to the applied pressure multiplied by the area of the bore will act on the surface of the liquid, forcing it down limb A. At the same time the liquid is forced up the bore of limb B until a state of equilibrium exists and the levels of the liquid stand at the same distance above and below the zero point. By taking into account the area of the tube bore and the density of the liquid it is possible to calculate the pressure from the difference in liquid levels, as the following example shows.

Let us assume that the manometer is of the mercury type having a bore area of $3\,\text{in}^2$, and that a pressure is applied to limb A such that at equilibrium the mercury levels are 4 in below and 4 in above zero. The difference in levels is H and its value is obtained by subtracting the lower level from the higher one; thus, $H = h_B - h_A = 4 - (-4) = 8\,\text{in}$.

Now, we must know the weight of the mercury column being supported, and this is calculated from volume multiplied by density. The volume in this case is $3H$ and the density of mercury is usually taken as $0.49\,\text{lb/in}^3$. Thus, the weight of the column is $3H \times 0.49 = 1.47 \times 8 = 11.76\,\text{lb}$, and as the pressure balancing this is weight divided by area, then $11.76/3 = 3.92\,\text{lb/in}^2$ is the pressure being applied to limb A and corresponding to a difference in mercury levels of 8 in. In the same manner, other pressures can be calculated from the corresponding values of level difference H.

In practice, manometers are used for checking the calibration of pressure gauges, and so it is usually more convenient to graduate the manometer scale directly in pounds per square inch. If 3·92 lb/in² is represented by 8 in, then, for the mercury manometer we have considered, 1 lb/in² is equal to 8/3·92, or 2·04 in, and so a scale can be graduated with marks spaced this distance apart, each representing an increment of 1 lb. The equivalent value 2·04 inHg to 1 lb is standard and results of calculations for different bore areas will show that they are independent of the areas.

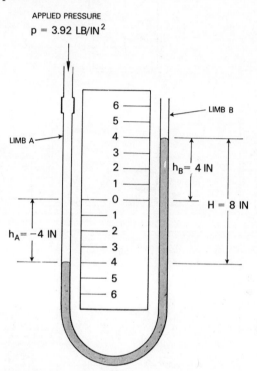

Fig 13.1 U-tube manometer

If water is used in the manometer the foregoing principles also apply, but as water has a much lower density than mercury, then for a given pressure the difference in level *H* for a water manometer will be much greater than that of a mercury manometer (2·04 inHg = 27·7 inH₂O very nearly).

Pressure/Weight Balancing

The measurement of pressure by balancing it against weights of known value is based on the principle of the hydraulic press, and as far as instruments are concerned, it finds a practical application in a hydraulic device known as the *dead-weight tester* and used for the calibration and testing of certain types of pressure gauge.

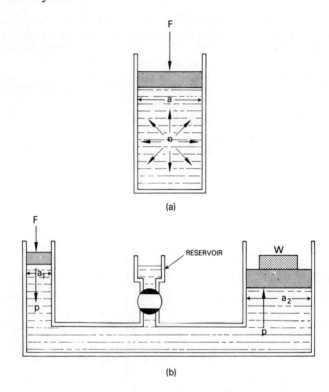
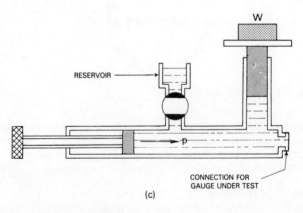

Fig 13.2 Pressure/weight balancing
 (a) Pressure produced in liquid
 (b) Hydraulic press
 (c) Dead-weight tester

Let us suppose that we have a cylinder containing a liquid as shown in Fig 13.2(a), and that a tight-fitting piston is placed on the liquid's surface. If now we try to push the piston down with a force F, we shall find that the piston will only be displaced by a very small amount, since the compressibility of liquids is very small. The pressure p produced in the liquid by pushing on the piston is equal to F/a and is transmitted to every part of the liquid and acts on all surfaces in contact with it.

In applying this principle to a hydraulic press we require essentially two interconnected cylinders as shown at (b), one of small cross-sectional area a_1, the other of large cross-sectional area a_2. Each cylinder is fitted with a piston and both are supplied with oil from a common reservoir. If a force F is exerted on the small piston then the additional pressure produced is $p = F/a_1$ and is transmitted throughout the liquid and therefore acts on the larger piston of area a_2. Thus, the force that can be exerted by this piston is equal to pa_2. If the press is designed to lift a weight W, then W will also be equal to pa_2. The weight that can be lifted by the application of a force F is multiplied in the ratio of the areas of the two pistons.

Figure 13.2(c) illustrates the hydraulic press principle applied to a dead-weight tester. When the piston in the horizontal cylinder is screwed in, a force is exerted and pressure is transmitted to the weighing piston in the vertical cylinder, so that it can be supported in a balanced condition by the oil column. In this application we are more interested in direct measurement of pressure and therefore need to know what weights are necessary to balance against required pressures. Now, the *area constant A* for a typical dead-weight tester is 0.125 in^2; thus, assuming that we require to balance a pressure p of 50 lb/in², then, from the relation $W = pA$ a weight of 6·25 lb is necessary. With this weight in position on the weighing piston the piston in the horizontal cylinder is screwed in until the weight is freely supported by the oil, which, at this point, is subjected to 50 lb/in². In practice, the weights are graded and are marked with the actual pressures against which they will balance.

ELASTIC PRESSURE-SENSING ELEMENTS

For pressure measurements in aircraft, it is obviously impracticable to equip the cockpit with U-tube manometers and dead-weight testers. It is the practice, therefore, to use *elastic pressure-sensing elements*, in which forces can be produced by applied pressures and made to actuate mechanical and/or electrical indicating elements. The sensing elements commonly used are Bourdon tubes, diaphragms, capsules and bellows.

Bourdon Tube

The Bourdon tube is about the oldest of the pressure sensing elements. It was developed and patented in 1850 by a Parisian watchmaker (whose name it bears) and has been in general use ever since, particularly in applications where the measurement of high pressure is necessary. The element, is essentially a length of metal tube, specially extruded to give it an elliptical cross-section, and shaped into the form of a letter C. The ratio between the major and minor axes depends on the sensitivity required, a larger ratio providing greater sensitivity. The material from which the tube is made may be either phosphor-bronze, beryllium-bronze or beryllium-copper. One end of the tube, the "free end," is sealed, while the other end is left open and fixed into a boss so that it may be connected to a source of pressure and form a closed system.

When pressure is applied to the interior of the tube there is a tendency for the tube to change from an elliptical cross-section to a circular one, and also to straighten out as it becomes more circular. In other words, it tends to assume its original shape. This is not such a simple process as it might appear and many theories have been advanced to explain it. However, a practical explanation sufficient for our purpose is as follows. Firstly, a tube of elliptical cross-section has a smaller volume than a circular one of the same length and perimeter. This being the case, an elliptical tube when connected to a pressure source is made to accommodate more of the liquid, or gas, than it can normally hold. In consequence, forces are set up which change the shape and thereby increase the volume. The second point concerns the straightening out of the tube as a result of its change in cross-section. Since the tube is formed in a C-shape then it can be considered as having an inner wall and an outer wall, and under "no pressure" conditions they are each at a definite radius from the centre of the C. When pressure is applied and the tube starts changing shape, the inner wall is forced towards the centre, decreasing the radius, and the outer wall is forced away from the centre thus increasing the radius. Now, along any section of the curved tube the effects of the changing radii are to compress the inner wall and to stretch the outer wall, but as the walls are joined as a common tube, reactions are set up opposite to the compressive and stretching forces so that a complete section is displaced from the centre of the C. Since this takes place at all sections along the tube and increases towards the more flexible portions, then the resultant of all the reactions will produce maximum displacement at the free end. Within close limits the change in angle subtended at the centre by a tube is proportional to the change of internal pressure, and within the limit of proportionality of the material employed, the displacement of the free end is proportional to the applied pressure.

The displacement of the free end is only small; therefore, in order to transmit this in terms of pressure, a quadrant and magnifying system is employed as the coupling element between tube and pointer.

Diaphragms

Diaphragms in the form of corrugated circular metal discs, owing to their sensitivity, are usually employed for the measurement of low pressures. They are always arranged so that they are exposed at one side to the pressure to be measured, their deflections being transmitted to pointer mechanisms, or as is more usual in aircraft applications, to a warning-light contact assembly. The materials used for their manufacture are generally the same as those used for Bourdon tubes. The purpose of the corrugations is to permit larger deflections, for given thicknesses, than would be obtained with a flat disc. Furthermore, their number and depth control the response and sensitivity characteristics; the greater the number and depth the more nearly linear is its deflection and the greater is its sensitivity.

Capsules

Capsules are made up of two diaphragms placed together and joined at their edges to form a chamber which may be completely sealed or open to a source of pressure. Like single diaphragms they are also employed for the measurement of low pressure, but they are more sensitive to small pressure changes. The operation of capsules in their various applications has already been described in the chapters on height and airspeed measuring instruments.

Bellows

A bellows type of element can be considered as an extension of the corrugated diaphragm principle, and in operation it bears some resemblance to a helical compression spring. It may be used for high, low or differential pressure measurement, and in some applications a spring may be employed (internally or externally) to increase what is termed the "spring-rate" and to assist a bellows to return to its natural length when pressure is removed.

The element is made from a length of seamless metal tube with suitable end fittings for connection to pressure sources or for hermetic sealing. Typical applications of bellows are described on pages 305 and 307.

DIRECT-READING PRESSURE GAUGES

These are almost entirely based on the Bourdon tube principle; two typical gauges are illustrated in Fig 13.3. The one at (*a*) is used in a hydraulic system, and as will be noted, it follows the same pattern as the gauge described on page 10.

The gauge shown at (*b*), which is employed in an aircraft brake system, consists of four Bourdon tube elements arranged side by side and operating individual pointers, which indicate the pressure available in the system and the pressure applied to each wheel brake.

REMOTE-INDICATING PRESSURE GAUGES

Gauges of this type are available in a variety of forms but all have one common feature; they consist of two main components, a transmitter unit located at the pressure source, and an indicator mounted on the appropriate panel. They have distinct advantages over direct-reading gauges; for example, the pressures of hazardous fluids such as fuel, engine oil and certain hydraulic fluids can be measured at their source and not brought up to the cockpit; also long pipelines are unnecessary thus saving weight.

The systems may be classified under two broad headings: (i) *liquid transmission*, and (ii) *electrical transmission*. In the first pressure is exerted on a special liquid and is measured in terms of its displacement, while in the other pressure is measured in terms of varying voltage and current signal combinations. In both cases elastic pressure-sensing elements are utilized.

Liquid Transmission System

The first gauge of this type ever to be introduced, and nowadays having only limited application, is the *transmitting capillary pressure gauge*, i.e. the transmitter unit and indicator are interconnected by a length of capillary tubing as shown in Fig 13.4.

The transmitter unit is in the form of a circular-shaped metal body containing a capsule so mounted that a small space is left all round it. The interior of the body is open to the source of pressure by means of a small port and

302 *Aircraft Instruments*

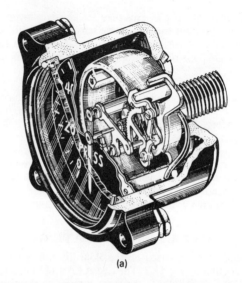

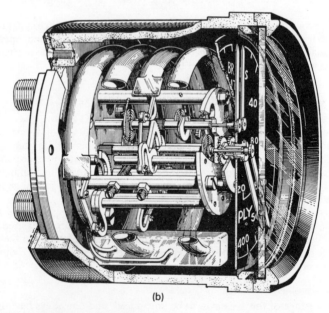

Fig 13.3 Direct-reading pressure gauges
(a) Single-element type
(b) Quadruple-element type

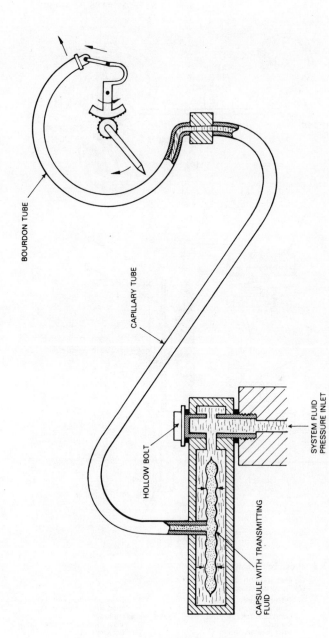

Fig 13.4 Liquid transmission type of pressure gauge

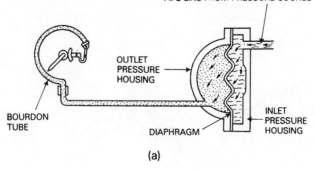

Fig 13.5 Pressure transmitter
  (a) Operating principle
  (b) Detailed sectional view

1 Inlet pressure housing
2 Centralizing disc
3 Centralizing knob
4 Locknut
5 Inlet pressure union
6 Non-return filler valve and plug cap
7 Mounting pillars and screws
8 Outlet pressure union
9 Diaphragm
10 Outlet pressure housing
11 Plug screw

hollow bolt assembly, and the capsule is connected by small-bore capillary tube to the Bourdon tube of the indicator. The length of the tubing is governed by the distance at which the pressure source is situated from the indicator location. The capsule, capillary and Bourdon tubes are completely filled with a liquid such as heptane (a paraffin hydrocarbon), thus constituting a sealed system.

When the capsule body is connected to the pressure source (an engine oil system for example) the oil passes through the hollow bolt and surrounds the capsule. As the oil is under pressure it tends to collapse the capsule and to displace the transmitting liquid throughout the system. Since the system is sealed, the liquid displacement causes the free end of the Bourdon tube to move and actuate the indicating element in the same manner as a direct-reading gauge.

The second type of gauge employing the liquid transmission principle is the one known as the *pressure transmitter* and shown in Fig 13.5. It is similar in many respects to the capillary type, but as the transmitter unit, tubing and gauge are all independent units, they are easily replaced should a failure occur in any of them.

The transmitter unit consist of two flanged circular-shaped aluminium-alloy castings and a Neoprene diaphragm. The diaphragm is interposed between the castings, which are bolted together to form two separate housings or chambers. The liquid whose pressure is to be measured is admitted to the smaller pressure-inlet housing, while the pressure-outlet housing is connected at the top by small-bore tubing to a direct-reading Bourdon-tube gauge, all three being completely filled after installation with a special low-viscosity mineral-base oil. A light-alloy disc, known as a centralizing disc, is provided in the inlet housing to ensure that the diaphragm remains in a central position and is not overdistended during filling or priming operations. It is positioned by the shaft located in the inlet housing. The lower filling connection contains a spring-loaded ball valve, while at the gauge connection a bleed valve is provided to expel any air bubbles that may collect during filling.

In operation, pressure is exerted on the diaphragm, which distends and displaces the transmitting fluid through the tubing and Bourdon tube in a similar manner to the transmitting capillary gauge system.

Electrical Transmission Systems

In the majority of cases these systems depend for the transmission of measured pressures on one or other of the synchronous systems already described in Chapter 10, and on the sensing of pressure by one or other of the elastic sensing elements. Thus, the transmitter units employed are made up of mechanical and electrical sections.

Figure 13.6 shows the arrangement of a transmitter working on the micro-Desynn principle. The pressure-sensing element consists of a bellows which is open to the pressure source. A cup-shaped pressing is fitted inside the bellows and forms a connection for a push-rod which bears against a rocking lever pivoted on a fixed part of the mechanism. A spring is provided inside the bellows.

The electrical element is of the same type as that shown in Fig 10.4 and is positioned in the transmitter body in such a manner that the eccentric pin is

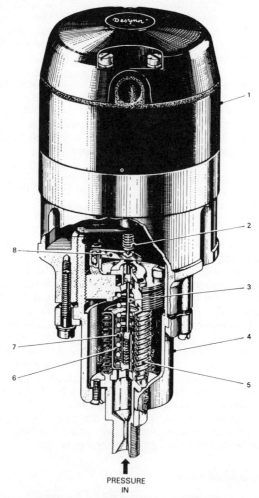

Fig 13.6 Micro-Desynn transmitter

1	Micro-Desynn transmitting element	5	Bellows
2	Eccentric pin	6	Cup-shaped pressing
3	Push rod	7	Spring
4	Pressure sensing element	8	Rocking lever

also in contact with the pivoted rocking lever. The indicator is of the normal Desynn system type.

When pressure is admitted to the interior of the bellows it expands and moves the push-rod up, thus rotating the rocking lever. This, in turn, moves the eccentric pin and brushes coupled to it through a small angle over the coils. The resistance changes produced set up varying voltage and current combinations within the indicator, which is calibrated for the appropriate pressure range.

Measurement of Pressure 307

RATIOMETER PRESSURE GAUGES

In addition to synchronous transmission types of pressure gauge, it is also possible to make use of gauges working on the ratiometer principle, and there are two versions in current use which we may consider. One is dependent on direct current for its operation and the other on a low-voltage alternating-current supply. Both systems require a transmitting unit consisting of mechanical and electrical sections.

The direct-current ratiometer pressure-gauge transmitter is a special adaptation of the micro-Desynn pressure transmitter described on page 305, the essential difference being in the electrical circuit arrangement. The element still has two brushes and resistance coils, but instead of the normal micro-Desynn method of connection (see also Fig 10.5) they are connected as a simple twin resistance parallel circuit. The two connections terminating at the coils are joined to the appropriate terminals of a ratiometer similar to that employed for temperature measurement. The operation is therefore quite simple; the movement of the bellows and brushes results in a change of circuit resistance proportional to the pressure change, which is measured as a coil current ratio.

The alternating-current type of ratiometer pressure gauge depends for its operation on the ratio of currents derived from variable inductor units. It is therefore a little more complicated than the one we have just considered.

A typical transmitter, shown in Fig 13.7, consists of a main body containing a bellows and two single-phase two-pole stators each surrounding a laminated salient-pole armature core. Both cores are on a common shaft and are so arranged that, as pressure increases, the lower core (A) moves further into its associated stator coil, while the upper core (B) moves further out of its coil. The coils are supplied with alternating current at 26 V, 400 Hz. The core poles are set 90° apart and the stators are also positioned so that the poles produced in them are at 90° to each other to prevent mutual magnetic interference. A spring provides a controlled loading on the bellows and armature cores, and is adjustable so as to set the starting position of the cores during calibration.

The essential parts of an indicator are illustrated in Fig 13.8. The coils around the laminated cores are connected to the transmitter stator coils, and as will be noted, a gap is provided in one limb of each core. The purpose of the gaps is to permit free rotation of two aluminium cam-shaped discs which form the moving element. The positioning of the discs on their common shaft is such that, when the element rotates in a clockwise direction (viewed from the front of the instrument in its normal position), the effective radius of the front disc a decreases in its airgap, while that of the rear disc b increases. The moving element is damped by a circular disc at the rear end of the shaft, which is free to rotate between the poles of a permanent magnet. A hairspring is provided to return the pointer to the off-scale position in the event of a power failure.

When the bellows expand under an increasing pressure, the armature cores move in their respective stators, and since the latter are supplied with alternating current, there is a change in the inductance of the coils. Thus core A, in moving further into its stator, increases the inductance and impedance, and core B, in moving out of its stator, decreases the inductance and impedance. The difference between the two may therefore be interpreted in terms of pressure.

As the stator coils are connected to the indicator coils in the form of a

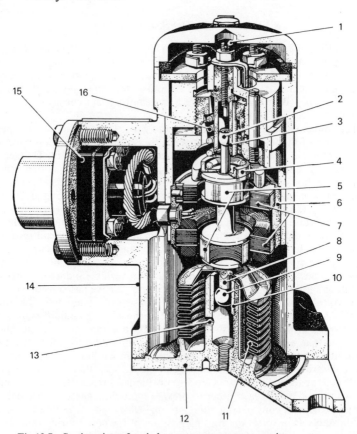

Fig 13.7 Section view of an inductor-type pressure transmitter

1. Overload stop screw
2. Centre spindle bearing
3. Guide bush
4. Aluminium cup
5. Armature cores
6. Aluminium housing
7. Stators and windings
8. Centre spindle assembly
9. Centre spindle bearing
10. Guide bush
11. Bellows
12. Base plate
13. Radial ducts
14. Body
15. Breeze plug assembly
16. Main spring

bridge network, then the changes in impedance will produce a change of current in the indicator coils at a predetermined ratio. The current is alternating, and so produces alternating fluxes in the laminated cores and across their gaps. It will be noted from Fig 13.8, that copper shading rings are provided at the airgaps. The effect of the alternating flux is to induce eddy currents in the rings, these currents in turn setting up their own fluxes which react with the airgap fluxes to exert a torque on the cam-shaped discs. The resulting movement of the cam discs is arranged to be in a direction determined by the coil carrying the greater current, and due to the disposition of the discs, this means there will be a difference between their torques. In the gap affected by the greater current the effective radius of its disc (a) decreases, thereby increasing the impedance and decreasing the torque, while in the gap affected by the weaker current, the

Measurement of Pressure 309

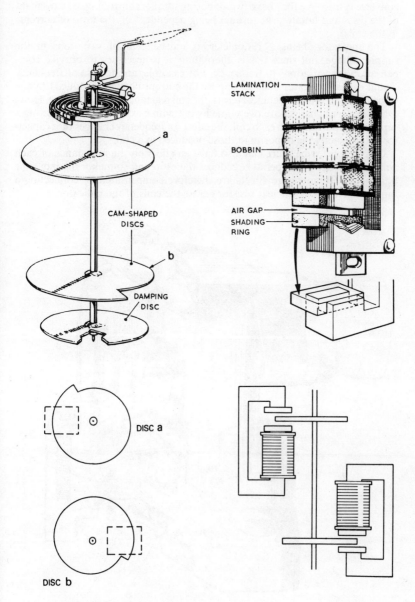

Fig 13.8 A.C. ratiometer elements

converse is true. We thus have two opposing torques controlling the movement of the discs and pointer, the torques being dependent on the ratio of currents in the coils.

The indicator, being a ratiometer, is independent of variations in the supply voltage, but since this is alternating, it is necessary to provide compensation for variations in frequency. For example, an increase of frequency would cause the stator coils to oppose the current changes produced by the transmitter, so that, in technical terms, the coil reactance would increase. However, reactance changes are overcome by the simple expedient of connecting a capacitor in parallel with each coil, the effects of frequency changes on a capacitor being exactly the opposite to those produced in a coil.

Changes in temperature can also have an effect on the impedance of each coil: an increase in temperature reduces the ratio and so makes the indicator under-read. Temperature effects are therefore compensated by connecting a high-temperature-coefficient resistor across the coils of the indicator.

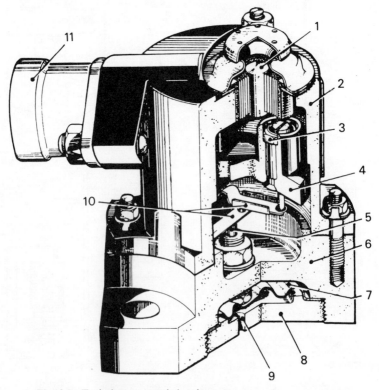

Fig 13.9 Typical pressure switch unit

1 Flame and water trap
2 Contact box
3 Contact adjuster
4 Terminal block
5 Push rod
6 Base
7 Diaphragm
8 Pressure plate
9 Pressure inlet
10 Contact arms
11 Plug

PRESSURE SWITCHES

In many of the aircraft systems in which pressure measurement is involved, it is necessary that a warning be given to pilots of either low or high pressures which might constitute hazardous operating conditions. In some systems also, the frequency of operation may be such that the use of a pressure-measuring instrument is not justified since it is only necessary for the pilot to know that an operating pressure has been attained for the period during which the system is in operation. To meet this requirement, pressure switches are installed in the relevant systems and are connected to warning or indicator lights located on the cockpit panels.

A typical switch is illustrated in Fig 13.9. It consists of a metal diaphragm bolted between the flanges of the two sections of the switch body. As may be seen, a chamber is formed on one side of the diaphragm and is open to the pressure source. On the other side of the diaphragm a push rod, working through a sealed guide, bears against contacts fitted in a terminal block connected to the warning or indicator light assembly. The contacts may be arranged to "make" on either decreasing or increasing pressure, and their gap settings may be preadjusted in accordance with the pressures at which warning or indication is required.

Pressure switches may also be applied to systems requiring that warning or indication be given of changes in pressure with respect to a certain datum pressure; in other words, as a differential pressure warning device. The construction and operation are basically the same as the standard type, with the exception that the diaphragm is subjected to a pressure on each side.

QUESTIONS

13.1 With the aid of diagrams explain how pressures are measured by: (*a*) a U-tube manometer, (*b*) a dead-weight tester.

13.2 (*a*) Describe the construction and operating principle of a Bourdon-tube pressure-sensing element.
(*b*) Name three other types of sensing element and state specific applications.

13.3 Is it possible to apply the Bourdon-tube principle to the measurement of negative pressure differential? (SLAET)

13.4 Describe a method of pressure measurement based on the principle of liquid transmission.

13.5 Describe how a d.c. type of ratiometer may be used in conjunction with a Desynn pressure transmitter.

13.6 Sketch a cross-section of a typical pressure switch; explain its operation.
(SLAET)

14 Measurement of Fuel Quantity and Fuel Flow

The measurement of the quantity of fuel in the tanks of an aircraft fuel system is an essential requirement, and in conjunction with measurements of the rate at which the fuel flows to the engine or engines, permits an aircraft to be flown at maximum efficiency compatible with its specified operating conditions. Furthermore, both measurements enable a pilot or engineer to quickly assess the remaining flight time and also to make comparisons between present engine performance and past or calculated performance.

Fuel-quantity indicating systems vary in operating principle and construction, the application of any one method being governed by the type of aircraft and its fuel system. Two principal methods currently applied utilize the principle of electrical signal transmission from units located inside the fuel tanks. In one method, mainly employed in the fuel systems of small and light aircraft, the tank units consist of a mechanical float assembly which controls an electrical resistance unit and varies the current flow to the indicating element. The second method, employed in high-performance aircraft fuel systems, measures fuel quantity in terms of electrical capacitance and provides a more highly accurate system of fuel gauging.

Fuel-flow measuring systems also vary in operating principle and construction but principally they consist of two units: a transmitter or meter, and an indicator. The transmitter is connected at the delivery side of the fuel system, and is an electro-mechanical device which produces an electrical output signal proportional to the flow rate which is indicated in either volumetric or mass units. In some systems an intermediate amplifier/computer is included to calculate a fuel-flow/time ratio and also to transmit signals to an indicator which presents integrated flow rate and fuel consumed information.

FLOAT-TYPE FUEL-QUANTITY INDICATING SYSTEMS

The components of a float-type system are shown schematically in Fig 14.1, together with the methods of transmitting electrical signals.

The float may be of cork specially treated to prevent fuel absorption, or it may be in the form of a lightweight metal cylinder suitably sealed. The float is attached to an arm pivoted to permit angular movement which is transmitted to an electrical element consisting of either a wiper arm and potentiometer, or a Desynn type of transmitter. As changes in fuel level take place the float arm moves through certain angles and positions the wiper arm or brushes to vary the resistance and flow of direct current to the indicator. As a result of the variations in current flow a moving coil or rotor within the indicator is deflected to position a pointer over the scale calibrated in gallons.

CAPACITANCE-TYPE FUEL-GAUGE SYSTEM

In its basic form, a capacitance-type fuel-gauge system consists of a variable capacitor located in the fuel tank, an amplifier and an indicator. The complete circuit forms an electrical bridge which is continuously being rebalanced as a result of differences between the capacitances of the tank capacitor and a reference capacitor. The signal produced is amplified to operate a motor, which positions a pointer to indicate the capacitance change of the tank capacitor and thus the change in fuel quantity.

Before going into the operating details of such a system, however, it is first necessary to discuss some of the fundamental principles of capacitance and its effects in electrical circuits.

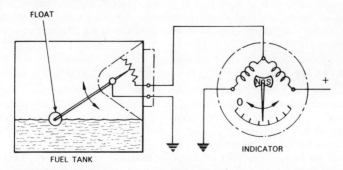

Fig 14.1 Simple float type of fuel quantity indicator

Electrical Capacitance

Whenever a potential difference is applied across two conducting surfaces separated by a non-conducting medium, called a dielectric, they have the property of storing an electric charge; this property is known as *capacitance*.

The flow of a momentary current into a capacitor establishes a potential difference across its plates. Since the dielectric contains no free electrons the current cannot flow through it, but the potential difference sets up a state of stress in the atoms comprising it. For example, in the circuit shown in Fig 14.2,

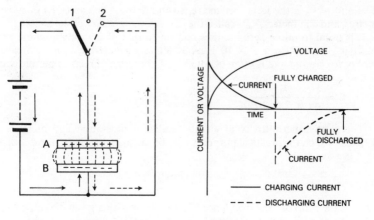

Fig 14.2 Charging and discharging of a capacitor

when the switch is placed in position 1 a rush of electrons, known as the charging current, takes place from plate A through the battery to plate B and ceases when the potential difference between the plates is equal to that of the battery.

When the switch is opened, the plates remain positively and negatively charged since the atoms of plate A have lost electrons while those at plate B have a surplus. Thus, electrical energy is stored in the electric field between the plates of the capacitor.

Placing the switch in position 2 causes the plates to be short-circuited and the surplus electrons at plate B rush back to plate A until the atoms of both plates are electrically neutral and no potential difference exists between them. This discharging current is in the reverse direction to the charging current; as shown in Fig 14.2.

Units of Capacitance

The capacitance or "electron-holding ability" of a capacitor is the ratio between the charge and the potential difference between the plates and is expressed in farads; one farad representing the ability of a capacitor to hold a charge of one coulomb (6.24×10^{18} electrons) which raises the potential difference between its plates by one volt.

Since the farad is generally too large for practical work, a sub-multiple of it is normally used called the microfarad ($1\mu F = 10^{-6} F$). In the application of the capacitor principle to fuel gauge systems, an even smaller unit, the picofarad ($1 pF = 10^{-12} F$) is the standard unit of measurement.

Factors on which Capacitance Depends

The capacitance of a parallel-plate capacitor depends on the area, a, of the plates, the distance, d, between the plates, and the capacitance, ϵ_a, of a unit cube of the dielectric material between the plates:

$$C = \epsilon_a \frac{a}{d} \quad \text{farads}$$

The unit of ϵ_a is the farad per metre, so that a must be expressed in square metres, and d in metres. ϵ_a is called the *absolute permittivity* of the dielectric.

It is usual to quote permittivities relative to that of a vacuum, whose permittivity, ϵ_0, is $1/4\pi \times 9 \times 10^9$ F/m. Relative permittivity, ϵ, is also called *dielectric constant* and is often denoted by K. In terms of relative permittivity,

$$C = \frac{K}{4\pi \times 9 \times 10^9} \frac{a}{d}$$

The relative permittivity of air at standard temperature and pressure is 1·00059, which for practical purposes may be taken as 1·0. Then, for example,

$$\frac{C_{water}}{C_{air}} = \frac{K_{water}}{K_{air}} = K_{water} \text{ very closely}$$

i.e. K is the ratio of the capacitance of a capacitor with a given dielectric to its capacitance with air between its plates.

The relative permittivities of some pertinent substances are as follows:

Air	1·00059
Water	81·07
Water vapour	1·007
Aviation gasolene	1·95
Aviation kerosene	2·10

Capacitors in Series and Parallel

The total capacitance of capacitors connected in series or parallel is obtained from formulae similar to those for calculating total resistance but which are applied in the opposite manner. Thus, for capacitors connected in series, the total capacitance is given by

$$\frac{1}{C_T} = \frac{1}{C_1} + \frac{1}{C_2} + \frac{1}{C_3} + \ldots$$

and for capacitors connected in parallel,

$$C_T = C_1 + C_2 + C_3 + \ldots$$

This is because the addition of capacitors in a circuit increases the plate area which, as already stated, is one of the factors on which capacitance depends.

Capacitors in Alternating-current Circuits

As already mentioned, when direct current is applied to a capacitor there is, apart from the initial charging current, no current flow through the capacitor. In applying the capacitance principle to fuel-gauge systems, however, a flow of current is necessary to make the indicator respond to the changes in capacitance arising from changes in fuel quantity. This is accomplished by supplying

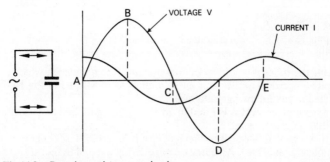

Fig 14.3 Capacitance in an a.c. circuit

the capacitance-type tank units with an alternating voltage, because whenever such voltage across a capacitor changes, electrons flow toward and away from it without crossing the plates and a resultant current flows which, at any instant, depends on the rate of change of voltage.

Figure 14.3 shows a capacitor connected to an alternating-current source. It will be observed from the graph that, as the voltage (V) rises rapidly at A a large current (I) will flow into the capacitor to charge it up. As the voltage increases towards B, however, the current decreases until at B, when the voltage is steady at some maximum value for a brief instant, the current has decreased to zero. From B to C the voltage decreases, the capacitor discharges and the current flows in the opposite direction, being a maximum at C, where the voltage is zero. From C to D the capacitor is charged in the opposite direction and the current flows in the same direction as the voltage but reaches zero at D, where the voltage is at some maximum value in the opposite direction. From D to E the capacitor again discharges. Thus, a charge and current flow in and out of the capacitor every half-cycle, the current leading the voltage by 90°.

The ratio between the voltage V and the current I is termed the *capacitive reactance*, meaning the opposition or resistance a capacitor offers to the flow of alternating current.

BASIC GAUGE SYSTEM

For fuel quantity measurement, the capacitors to be installed in the tanks must differ in construction from those normally employed in electrical equipment. The plates therefore take the form of two tubes mounted concentrically with a narrow air space between them, and extending the full depth of a fuel tank. Constructed in this manner, two of the factors on which capacitance depends are fixed, while the third factor, dielectric constant, is variable since the medium

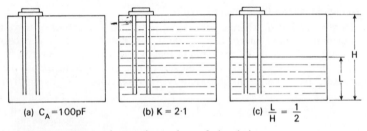

Fig 14.4 Changes in capacitance due to fuel and air

between the tubes is made up of fuel and air. The manner in which changes in capacitance due to fuel and air take place is illustrated in Fig 14.4 and is described in the following paragraphs.

At (a) a tank capacitor is fitted in an empty fuel tank and its capacitance in air is 100 pF, represented by C_A.

At (b) the tank is filled with a fuel having a K value of 2·1, so that the capacitor is completely immersed. As stated on page 314, K is equal to the ratio of capacitance using a given dielectric (in this case C_T) to that using air; therefore,

$$K = \frac{C_T}{C_A} \tag{1}$$

From eqn (1) $C_T = C_A K$, and it is thus clear that the capacitance of the tank unit at (b) is equal to 100 × 2·1, i.e. 210 pF. The increase of 110 pF is the added capacitance due to the fuel and may be represented by C_F. The tank unit may therefore be represented electrically by two capacitors in parallel and of a total capacitance

$$C_T = C_A + C_F \tag{2}$$

In Fig 14.4(c), the tank is only half full and so the total capacitance is $100 + 55$, or 155 pF. The added capacitance due to fuel is determined as follows. By transposing eqn (2), $C_F = C_T - C_A$, and by substituting $C_A K$ for C_T we obtain $C_F = C_A K - C_A$, which may be simplified as

$$C_F = C_A(K - 1) \tag{3}$$

the factor $(K - 1)$ being the increase in the K value over that of air.

Now, the fraction of the total possible fuel quantity in a "linear tank" at any given level is given by L/H, where L is the height of the fuel level and H the total height of the tank. Thus by adding L/H to eqn (3) the complete formula becomes

$$C_F = \frac{L}{H}(K - 1)C_A \tag{4}$$

The circuit of a basic gauge system is shown in Fig 14.5. It is divided into two sections or loops by a resistance R, both loops being connected to the secondary winding of a power transformer. Loop A contains the tank capacitor C_T and may therefore be considered as the sensing loop of the bridge since it detects current changes due to changes in capacitance. Sensing loop voltage V_S remains constant.

Loop B, which may be considered as the balancing loop of the bridge, contains a reference capacitor C_R of fixed value, and is connected to the transformer via the wiper of a balance potentiometer so that the voltage V_B is variable.

The balance potentiometer is contained within the indicator together with a two-phase motor which drives the potentiometer wiper and indicator pointer. The reference phase of the motor is continuously energized by the power transformer and the control phase is connected to the amplifier and is only energized when an unbalanced condition exists in the bridge.

The amplifier, which may be of either the thermionic-valve or the transistor type, has two main stages: one for amplifying the signal produced by bridge unbalance, and the other for discriminating the phase of the signal which is then supplied to the motor.

Let us consider the operation of the complete circuit when fuel is being drawn off from a full tank. Initially, and at the constant full-tank level, the sensing current I_S is equal to the balancing current I_B; the bridge is thus in balance and no signal voltage is produced across R.

As the fuel level drops, the tank capacitor has less fuel around it; therefore the added capacitance (C_F) has decreased. The tank unit capacitance decreases and so does the sensing current I_S, the latter creating an unbalanced bridge condition with balancing current I_B predominating through R.

A signal voltage proportional to $I_B R$ is developed across R and is amplified and its phase detected before being applied to the control phase of the indicator motor. The output signal is a half-wave pulse, a feature of both valves and transistors in discriminator circuits, and in order to convert it into a full-wave signal, a capacitor is connected in parallel with the control winding. A

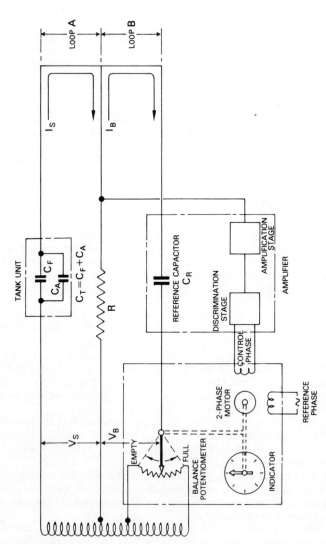

Fig 14.5 Circuit of a basic fuel-gauge system

capacitor is also connected in series with the reference winding to form what is termed a series-resonant circuit. This circuit ensures that the currents in both phases are 90° out of phase, the current in the control phase either leading or lagging the reference phase depending on which loop of the bridge circuit is predominating.

In the condition we are considering, the balancing current is predominating; therefore, the control-phase current lags behind that of the reference phase causing the motor and balance potentiometer wiper to be driven in such a direction as to decrease the balancing current I_B.

When the current I_B equals the current I_S, the bridge is once again in balance, the motor stops rotating and the indicator pointer registers the new, lower value.

Effects of Fuel Temperature Changes

With changes in temperature the volume, density and relative permittivity of fuels are affected to approximately the same degree as shown in Fig 14.6, which is a graph of the approximate changes occurring in a given mass of fuel. From this it should be noted that $K - 1$ is plotted, since for a gauge system

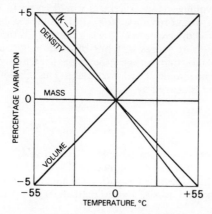

Fig 14.6 Temperature effects on fuel characteristics

measuring fuel quantity by volume, the indicator pointer movement is directly dependent on this. It should also be noted that, although it varies in the same way as density, the percentage change is greater.

Thus a volumetric gauge system will be subject to a small error due to variations in fuel temperature. Furthermore, changes in K and density also occur in different types of fuel having the same temperature. For example, a gauge system which is calibrated for a K-value of 2·1 has a calibration factor of $2·1 - 1 = 1·1$. If the same system is used for measuring a quantity of fuel having a K-value of 2·3, then the calibration factor will have increased to 1·3 and the error in indication will be approximately

$$\frac{1·3}{1·1} \times 100 - 100\% = 18\%$$

i.e. the gauge would overread by 18%.

MEASUREMENT OF FUEL QUANTITY BY WEIGHT

A more useful and accurate method of measuring fuel quantity is to do so in terms of its mass or weight. This is because the total power developed by an engine, or the work it performs during flight, depends not on the volume of fuel but on the energy it contains, i.e. the number of molecules that can combine with oxygen in the engine. Since each fuel molecule has some weight and also because one pound of fuel has the same number of molecules regardless of temperature and therefore volume, the total number of molecules (total available energy) is best indicated by measuring the total fuel weight.

In order to do this, the volume and density of the fuel must be known and the product of the two determined. The measuring device must therefore be sensitive to both changes in volume and density so as to eliminate the undesirable effects due to temperature.

This will be apparent by considering the example of a tank holding 1,000 gal of fuel having a density of 6 lb/gal at normal temperature. Measuring this volumetrically we should of course obtain a reading of 1,000 gal, and from a mass measurement, 6,000 lb. If a temperature rise should increase the volume by 10%, then the volumetric measurement would go up to 1,100 gal, but the mass measurement would remain at 6,000 lb because the density of the fuel (weight/volume) would have decreased when the temperature increased.

For the calibration of gauges in terms of mass of fuel, the assumption is made that the relationship between the relative permittivity (K) and the density (ρ) of a given sample of fuel is constant. This relationship is called the *capacitive index* and is defined by

$$\frac{K-1}{\rho}$$

A gauge system calibrated to this expression is still subject to indication errors, but they are very much reduced. This may be illustrated by a second example.

Assuming that the system is measuring the quantity of a fuel of nominal $K = 2 \cdot 1$ and of nominal density $\rho = 0 \cdot 779$, then its capacitive index is

$$\frac{2 \cdot 1 - 1}{0 \cdot 779} = 1 \cdot 412$$

Now, if the same gauge system measures the quantity of another fuel for which the nominal K and ρ values are respectively 2·3 and 0·85, then its capacitive index will increase thus:

$$\frac{2 \cdot 3 - 1}{0 \cdot 85} = 1 \cdot 529$$

However, the percentage error is now

$$\frac{1 \cdot 529}{1 \cdot 412} \times 100 - 100\% = 8\% \text{ approx.}$$

and this is the amount by which the gauge would over-read.

COMPENSATED GAUGE SYSTEMS

Although the assumed constant permittivity and density relationship results in a reduction of the indication error, tests on properties of fuels have shown that, while the capacitive index varies from one fuel to another, this variation tends to follow the permittivity. Thus, if a gauge system can also detect changes in the permittivity of a fuel as it departs from its nominal value, then the density may be inferred to a greater accuracy, resulting in an even greater reduction of indication errors.

The gauge systems now in use are therefore of the permittivity or inferred-density compensated type, the compensation being effected by a reference capacitor added to the balance loop of the measuring circuit and in parallel with the capacitor C_R, as shown in Fig 14.7.

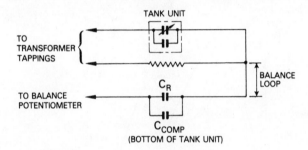

Fig 14.7   Circuit connection of compensator capacitor

A compensator is similar in construction to a standard tank unit and is usually fitted to the bottom of a unit to ensure that the compensator is always immersed in fuel. Located in this manner, its capacitance is determined solely by the permittivity, or K-value, of the fuel, and not by its quantity as in the case of the tank unit. In addition to variable voltage, the balancing loop will also be subjected to variable capacitance, which means that balancing current I_B will be affected by variations in K as well as sensing-loop current I_S.

Let us assume that the bridge circuit (see Fig 14.5) is in balance and that a change in temperature of the fuel causes its K-value to increase. The tank unit capacitance will increase and so current I_S will predominate to unbalance the bridge and to send a signal voltage to the amplifier and control phase of the motor. This signal will be of such a value and phase that an increase in balancing loop current is required to balance the bridge, and so the motor must drive the wiper of the balance potentiometer to decrease the resistance. Since this is in the direction towards the "tank full" condition, the indicator will obviously register an increase in fuel quantity. The increase in K, however, also increases the compensator capacitance so that balancing-loop current I_B is increased simultaneously with, but in opposition to, the increase of sensing-loop current I_S. A balanced bridge condition is therefore obtained which is independent of the balance potentiometer.

In practice, there is still an indication error due to the fact that the density

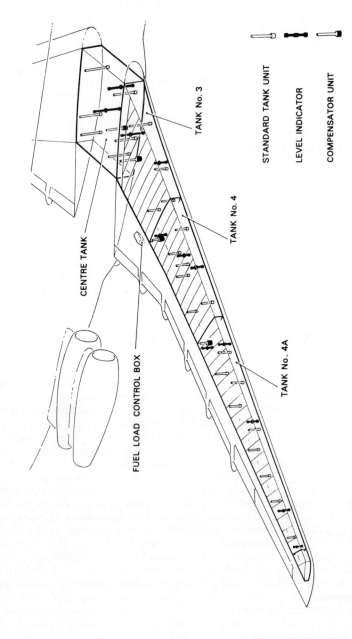

Fig 14.8 Location of tank units

also varies with temperature, and this is not directly measured. But the percentage increase of density is not as great as that of $K - 1$, and so by careful selection of the compensator capacitance values in conjunction with the reference capacitance, the greatest reduction in overall gauge error is produced.

LOCATION AND CONNECTION OF TANK UNITS

In practical capacitance-type fuel-quantity indicating systems, a number of tank units are disposed within the fuel tanks as illustrated in Fig 14.8, and are connected in parallel to their respective indicators. The reasons for this are to ensure that indications remain the same regardless of the attitude of an aircraft and its tanks and also of any surging of the fuel. This may be understood by

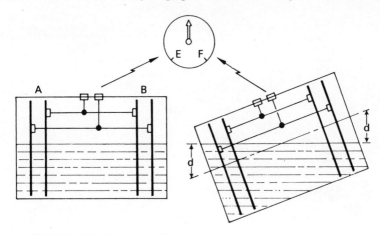

Fig 14.9   Attitude compensation

considering a two-unit system as shown in Fig 14.9. If the tank is half-full and in a level attitude, each tank unit will have a capacitance of half its maximum value, and since they are connected in parallel the total capacitance measured will produce a "half-full" indication. When the tank is tilted, and because the fuel level remains the same, unit A is immersed deeper in the fuel by the distance d and gains some capacitance, tending to make the indicator overread. Tank unit B, however, has moved out of the fuel by the same distance d and loses an equal amount of capacitance. Thus, the total capacitance remains the same as for the level-tank attitude and the indication is unchanged.

In the majority of cases, tank units are designed for internal mounting, their connections being brought out through terminals in the walls of the tanks. For some aircraft tank systems flange-mounted units may also be provided.

A typical standard tank unit and a compensator unit are shown at (a) and (b) respectively of Fig 14.10. The tubes are of aluminium alloy held apart by pairs of insulating cross-pins. Electrical connections are made through coaxial connectors mounted on a bracket attached to a nylon sleeve secured to the upper part of the outer tube. Two further nylon sleeves, one at each end, are secured to the outer tube, and to each sleeve a rubber ring is attached. The purpose of the rubber ring is to hold the unit in position at supporting fixtures within the tank. The reference unit of the compensator consists of three concentric tubes held apart by insulating buttons; the outer of the three tubes acts as an earth screen.

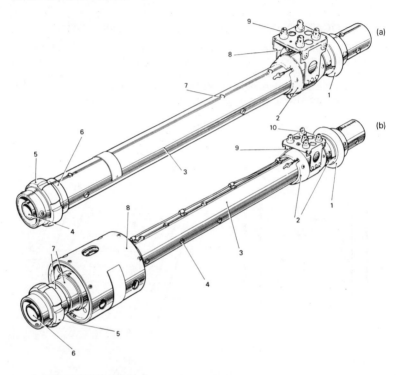

Fig 14.10 Tank units
(a) Standard unit
1 Rubber ring
2 Nylon sleeve
3 Outer tube
4 Inner tube
5 Rubber ring
6 Nylon sleeve
7 Insulating cross-pin
8 Bracket
9 Miniature coaxial connector

(b) Compensator unit
1 Rubber ring
2 Nylon sleeves
3 Outer tube
4 Insulating cross-pin
5 Rubber ring
6 Inner tube
7 Nylon sleeves
8 Reference unit
9 Bracket
10 Miniature coaxial connector

Characterized Tank Units

The fuel tanks of an aircraft may be separate units designed for installation in wings and in centre sections, or they may form an integral sealed section of these parts of the structure. This means, therefore, that tanks must vary in contour to suit their chosen locations, with the result that the fuel level is established from varying datum points.

Figure 14.11 represents the contour of a tank located in an aircraft wing, and as will be noted the levels of fuel from points A, B and C are not the same. When standard tank units are positioned at these points the total capacitance will be the sum of three different values due to fuel (C_F), and as the units produce the same change of capacitance for each inch of wetted length, the indicator scales will be non-linear corresponding to the non-linear characteristic of the tank contour.

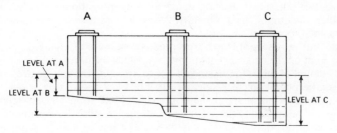

Fig 14.11   Non-linear fuel tank

The non-linear variations in fuel level are unavoidable, but the effects on the graduation spacing of the indicator scale can be overcome by designing tank units which measure capacitance changes proportional to tank contour. The non-linear tank units so designed are called *characterized tank units*, the required effect being achieved either by altering the diameter of the centre electrode or by varying the area of its conducting surface at various points over its length, to suit the tank contour.

Empty and Full Position Adjustments

In order that an indicator pointer shall operate throughout its range between the two principal datums corresponding to empty- and full-tank conditions, it is necessary during calibration to balance the current and voltage of the circuit sensing and balancing loops at these datums. This is achieved by connecting two manually controlled potentiometers into the circuit as shown in Fig 14.12.

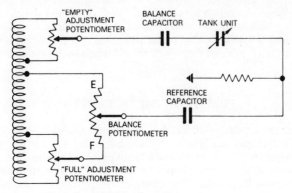

Fig 14.12   "Empty" and "full" adjustments

The "empty" potentiometer is connected at each end to the supply transformer and its wiper is connected to the tank units via a balance capacitor. When a tank is empty, due to the "empty" capacitance of the tank units, current will still flow through them. The balance potentiometer wiper will also be at its "empty" position, but since it is earthed at this point, no current will flow through the reference capacitor. However, current does flow through the balance capacitor and it is the function of the "empty" potentiometer to balance this out. The balancing signal from the potentiometer is fed to the amplifier, the output signal of which drives the indicator and servomotor to the "empty" position.

The "full" adjustment may be regarded as a means of changing the position of the point on the balance potentiometer at which the balance voltage for any given amount of fuel is found, and also of determining the voltage drop across the balance potentiometer. Reference to Fig 14.12 shows that, if the "full" potentiometer wiper is set at the bottom, the "full" transformer secondary voltage will be applied to the balance potentiometer. With the wiper at the top, resistance is introduced into the circuit so that a smaller voltage is applied to the balance potentiometer. Therefore, the distance the balance potentiometer wiper needs to move to develop a given balance voltage can be varied.

Fail-safe and Test Circuits

In the event of failure of the output from an amplifier there can be no response to error and balancing signals, and to ensure that a pilot or engineer is not misled by an indicator remaining stationary at some fuel quantity value, a fail-safe circuit or a test circuit is incorporated in a gauge system.

A typical fail-safe circuit consists of a capacitor and a resistor, connected in the indicator motor circuit so that the reference winding supply is paralleled to the control winding. A small leading current always flows through the parallel circuit, but under normal operating conditions of the system, it is suppressed by the tank unit signals flowing through the motor control winding. When these signals cease due to a failure, the current in the parallel circuit predominates and flows through the control winding to drive the pointer slowly downwards to the "empty" position.

A test circuit incorporates a switch mounted adjacent to its appropriate indicator. When the switch is held in the "test" position a signal simulating an emptying tank condition is introduced into the indicator motor control winding, causing it to drive the pointer towards zero if the circuit is operable. When the switch is released the pointer should return to its original indication.

FUEL FLOW MEASUREMENT

In analysing current designs of fuel-flow measuring systems it will be found that they come within two main groups: (i) *independent fuel flow*, and (ii) *integrated*. There are a variety of types in use and it is not possible to go into the operating details of them all. However, a system which may be considered representative of each group has been selected to illustrate the applications of fundamental requirements and principles.

Independent Fuel Flow System

This system consists of a transmitter and indicator and requires 28 V direct current for its operation.

The transmitter, shown in Fig 14.12, has a cast body with inlet and outlet connections in communication with a spiral-shaped metering chamber containing the metering assembly. The latter consists of a metering vane pivoted so that it can be angularly displaced under the influence of fuel passing through the chamber. A small gap is formed between the edge of the vane and the

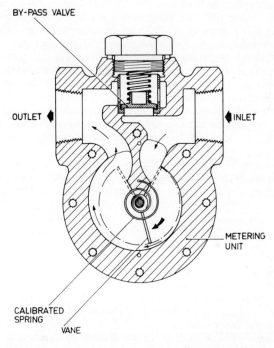

Fig 14.13 Section view of a rotating-vane fuel-flow transmitter

chamber wall, which, on account of the volute form of the chamber, increases in area as the vane is displaced from its zero position. The variation in gap area controls the rate of vane displacement which is faster at the lower flow rates (gap narrower) than at the higher ones. Thus, its function may be likened to an airspeed-indicator square-law compensator which it will be recalled is a device for opening up an indicator scale. The vane is mounted on a shaft carried in two bushed plain bearings, one in each cover plate enclosing the metering chamber.

At one end, the shaft protrudes through its bearing and carries a two-pole ring-type magnet which forms part of a magnetic coupling between the vane and the electrical transmitting unit. In this particular system the unit is a precision potentiometer; in some designs an a.c. synchor may be used. The shaft of the potentiometer (or synchro) carries a two-pole bar-type magnet which is located inside the ring magnet. The interaction of the two fields provides a "magnetic lock" so that the potentiometer wiper (or synchro rotor) can follow any angular displacement of the metering vane free of friction.

The other end of the metering vane shaft also protrudes through its bearing and carries the attachment for the inner end of a specially calibrated control spring. The outer end of the spring is anchored to a disc plate which can be rotated by a pinion meshing with teeth cut in the periphery of the plate. This provides for adjustment of the spring torque during transmitter calibration.

Any tendency for the metering assembly and transmission element to oscillate under static flow rate conditions is overcome by a liquid damping system, the liquid being the fuel itself. The system comprises a damping chamber containing a counterweight and circular vane which are secured to the same end of the metering vane shaft as the control spring. The damping chamber is secured at one side of the transmitter body, and except for a small bleed hole in a circular blanking plate, is separated from the metering chamber. The purpose of the bleed hole is, of course, to permit fuel to fill the damping chamber and thus completely immerse the counterweight assembly. The effectiveness of the damping system is uninfluenced by the fuel flow. A threaded plug in the outer cover of the damping chamber provides for draining of fuel from the chamber.

The indicator is of simple construction, being made up of a moving-coil milliammeter which carries a single pointer operating over a scale calibrated in gallons per hour, pounds per hour or kilogrammes per hour. The signals to the milliammeter are transmitted via a transistorized amplifier which is also contained within the indicator case. In systems employing synchronous transmission, the indicator pointer is operated by the rotor of a receiver synchro.

Operation

When fuel commences to flow through the main supply line it enters the body of the transmitter and passes through the metering chamber. In doing so it deflects the metering vane from its zero position and tends to carry it round the chamber. Since the vane is coupled to the calibrated spring, the latter will oppose movement of the vane, permitting it only to take up an angular position at which the tension of the spring is in equilibrium with the rate of fuel flow at any instant. Through the medium of the magnetic-lock coupling the vane will also cause the potentiometer wiper to be displaced, and with a steady direct voltage across the potentiometer the voltage at the wiper is directly proportional to the fuel flow. The voltage is fed to the amplifier, whose output current drives the milliammeter pointer to indicate the fuel flow.

In a system employing synchros, the current flow due to differences in angular position of the rotors will drive the indicator synchro rotor directly to the null position and thereby make the indicator pointer read the fuel flow.

In the type of transmitter considered it is also necessary to provide a by-pass for the fuel in the event of jamming of the vane or some other obstruction causing a build-up of pressure on the inlet side. It will be noted from Fig 14.13 that the valve is of the simple spring-loaded type incorporated in the metering chamber. The spring tension is adjusted so that the valve lifts from its seat and allows fuel to by-pass the metering chamber when the difference of pressure across the chamber exceeds $2 \cdot 5 \text{lb/in.}^2$

INTEGRATED FLOWMETER SYSTEM

We may broadly define an *integrated flowmeter* system as one in which the element indicating fuel consumed is combined with that required for fuel flow, thus permitting the display of both quantities in a single instrument.

In order to accomplish this it is also necessary to include in the system a device which will give directly the fuel consumed over a period of time from the flow rate during that period. In other words, a time integrator is needed to work out fuel consumed in the ratio of fuel flow rate to time.

Such a device may be mechanical, forming an integral part of an indicator mechanism, or as in electronic flowmeter systems it may be a special dividing stage within the amplifier or even a completely separate integrator unit. A typical system to which principles of integration are applied will now be studied.

The system consists of three principal units: flow transmitter, electronic relay or computer, and indicator. Its operation depends on the principle that the torque required to accelerate a fluid to a given angular velocity is a measure of the fluid's mass flow rate. The angular velocity, which is imparted by means of a rotating impeller and drum, sets up a reaction to establish relative angular displacements between the impeller and drum. Inductive-type pick-offs sense the displacements in terms of signal pulses proportional to the flow rate and supply them via the amplifier/computer, to the indicator.

The transmitter, shown schematically sectioned in Fig 14.14, consists of a light-alloy body containing a flow metering chamber, a motor-driven impeller

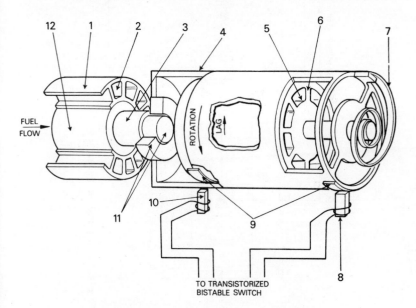

Fig 14.14 Fuel-flow transmitter of a typical integrated system

1 Turbine
2 Fluid passage
3 Shaft
4 Light-alloy body
5 Fluid passage
6 Impeller
7 Restraining spring
8 Pick-off assembly
9 Magnets
10 Pick-off assembly
11 Magnetic coupling
12 Rotor

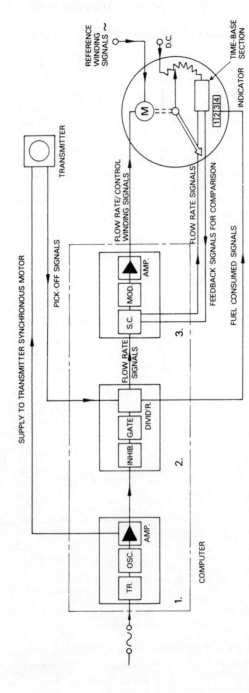

Fig 14.15 Integrated fuel flowmeter system

assembly, and an externally mounted inductor coil assembly. The impeller assembly consists of an outer drum which is driven through a magnetic coupling and reduction gear, by a synchronous motor, and an impeller incorporating vanes to impart angular velocity to fuel flowing through the metering chamber. The drum and impeller are coupled to each other by a calibrated linear spring. The motor is contained within a fixed drum at the inlet end and rotates the impeller at a constant speed. Straightening vanes are provided in the fixed drum to remove any angular velocity already present in the fuel before it passes through the impeller assembly. A point to note about the use of a magnetic coupling between the motor and impeller assembly is that it overcomes the disadvantages which in this application would be associated with rotating seals. The motor and its driving gear are isolated from fuel by enclosing them in a chamber which is evacuated and filled with an inert gas before sealing.

Each of the two pick-off assemblies consist of a magnet and an iron-cored inductor. One magnet is fitted to the outer drum while the other is fitted to the impeller, thus providing the required angular reference points. The magnets are so positioned that under zero flow conditions they are effectively in alignment with each other. The coils are located in an electrical compartment on the outside of the transmitter body, together with transistorized units which amplify and switch the signals induced.

The computer performs the overall function of providing the power for the various circuits of the system, detecting the number of impulses produced at the transmitter, and computing and integrating the fuel flow rate and amount of fuel consumed. It consists of a number of stages interconnected in three distinct sections as shown in the block diagram of Fig 14.15. The stages of sections 2 and 3 consist of transistors and their associated coupling capacitors and resistors. The power supply section (1) controls the voltage and frequency of the supply to the transmitter synchronous motor, and consists of a transformer, transistorized crystal oscillator, output and power amplifier units.

From the diagram it will be noted that section 2 is made up of three stages: inhibitor, gate and divider. The respective functions of these stages are: to suppress all transmitter signals below a certain flow rate; to control or gate the pulse signals from the power-supply oscillator, and to produce output signals proportional to true flow rate; to provide the time dividing factor and output pulses representing unit mass of fuel consumed. Section 3 is also made up of three stages; signal comparator, modulator and servo amplifier. The respective functions of these stages are: to compare the transmitter output signals with time-base signals fed back from the indicator; to combine the comparator output with 400 Hz alternating current and produce a new output; to provide an operating signal to the indicator servomotor control winding.

The indicator employs a flow indicating section consisting of a 400 Hz servomotor which drives a pointer and potentiometer wiper through a reduction gear train. The reference winding of the motor is supplied with a constant alternating voltage, while the control winding receives its signals from the computer servo amplifier. The potentiometer is supplied with direct current and its wiper is electrically connected to a transistorized time-base section, also within the indicator. Transmitter output signals are also fed into the time-base section via a pre-set potentiometer which forms part of the computer signal comparator stage. The difference between the time-base and the indicated fuel-flow signal voltages is fed to the servomotor which operates to reduce the error voltage to zero and so to correct the indicated fuel flow.

The fuel-consumed section of the indicator consists of a solenoid-actuated 5-drum digital counter and a pulse amplifier. The amplifier receives a pulse from the divider stage of the computer for each unit mass of fuel consumed and feeds its output to the solenoid, which advances the counter drums appropriately. A mechanical reset button is provided for resetting the counter to zero.

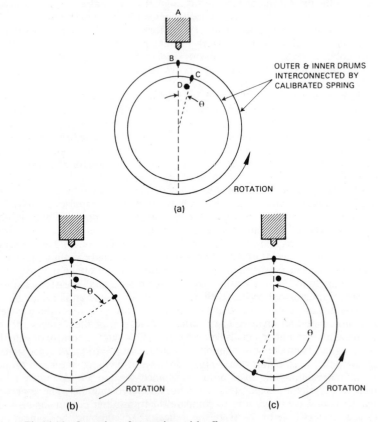

Fig 14.16 Operation of transmitter pick-offs

(a) Zero fuel flow
    A   Two pick-off coils (one behind the other)
    B, C Magnets
    D   Stop (gives $3°$ to $5°$ deflection)
    θ Lag angle at which both drums rotate together
(b) Cruising fuel flow
(c) Maximum fuel flow

Operation

When electrical power is switched on to the system, the synchronous motor in the transmitter is operated to drive the impeller assembly at a constant speed. Under zero fuel flow conditions the magnets of the pick-off assemblies are effectively in line with one another, although in practice there is a small angular difference established to maintain a deflection representing a specific minimum flow rate. This is indicated in Fig 14.16(a). As the fuel flows through the transmitter metering chamber, a constant angular velocity is imparted to the fuel by the rotating impeller and drum assembly, and since the two are interconnected by a calibrated spring, a reaction torque is created which alters the angular displacement between impeller and drum, and their corresponding magnets. Thus angular displacement is proportional to flow rate. Figs 14.16(b) and (c) illustrate the displacement for typical cruising and maximum fuel flow rates.

The position of each magnet is sensed by its own pick-off coil, and the primary pulses induced as each magnet moves past its coil are fed to the dividing stage in the computer (see also Fig 14.15). The output from this stage is fed to the control winding of the indicator servomotor via section 3 of the computer, and the indicator pointer is driven to indicate the fuel flow. At the same time, the motor drives the potentiometer wiper, producing a signal which is fed back to the signal comparator stage and compared with the output produced by the transmitter. Any resultant difference signal is amplified, modulated and power amplified to drive the indicator motor and pointer to a position indicating the actual fuel flow rate.

The computer divider stage also uses the transmitter signals to produce pulse "time" signals for the operation of the fuel-consumed counter of the indicator. During each successive revolution of the transmitter impeller assembly the pulses are added and divided by a selected ratio, and then supplied to the counter as an impulse for each kilogramme or pound of fuel consumed.

QUESTIONS

14.1 Describe the construction and operation of a float type of fuel-quantity gauge with which you are familiar.

14.2 Explain the operating principle of a capacitor and state the factors on which its capacitance depends.

14.3 Define (a) the units in which capacitance is expressed, (b) permittivity.

14.4 (a) Explain how alternating current applied to a capacitor causes current to flow across it.
(b) What is the term used to express the ratio of voltage to current?

14.5 Give the formula for calculating the total capacitance of a circuit containing (a) capacitors in series, (b) capacitors in parallel.

14.6 (a) Which of the capacitances variables is utilized in a capacitor-type fuel-gauge system?
(b) Describe the construction of a typical tank unit.

14.7 Draw a circuit of a typical capacitor-type fuel-quantity indicating system. Explain the operating principle. (SLAET)

14.8 (a) What effects do temperature changes have on the fuels used?
(b) Explain how these are compensated in a fuel-quantity indicating system.

14.9 Explain why it is necessary to install a number of tank units in a fuel tank.

14.10 Why is it preferable for fuel-quantity indicating systems to measure fuel weight rather than fuel volume?

14.11 What adjustments are normally provided in a capacitance-type fuel-quantity indicating system?

14.12 Explain the function of the test switch incorporated in some quantity fuel-indicating systems.

14.13 Describe the construction of a fuel flowmeter indicator and explain the basic principle of operation. (SLAET)

14.14 What is the function of the by-pass valve as fitted to certain types of fuel flowmeter?

14.15 (a) What is an integrated flowmeter system?
(b) Describe a method of achieving integration.

15 Engine Power and Control Instruments

Power of piston engines, turbojet and turboprop engines, refers to the amount of thrust available for propulsion and is expressed as power ratings in units of brake or shaft horsepower (bhp or shp) at a propeller shaft or in pounds of thrust at the jet pipe.

The power ratings of each of the various types of engine are determined during the test-bed calibration runs conducted by the manufacturer for various operating conditions such as take-off, climb and normal cruising. For example, the rated thrust of a turbojet engine at take-off may be limited to five minutes at a turbine temperature of 795°C. This does not mean, of course, that the engine is suddenly going to disintegrate should these limitations be exceeded, but frequent excesses will obviously set up undesirable stresses leading to deterioration of vital internal parts of the engine.

All ratings are therefore established with a view to conserving the lives of engines and contributing to longer periods between overhauls. It is thus necessary for engines to be instrumented for the indication of power being developed. The instruments associated with power indication for the various types of engine are listed in Table 15.1.

Table 15.1 Power indication instruments

Indicator	Type of engine				
	Piston		Turbojet		Turboprop
	Unsupercharged	Supercharged	Centrifugal compressor	Axial flow compressor	
Tachometer	X	X	X	X	X
Fuel flow		X	X	X	
Manifold pressure		X			
Torque pressure		X			X
Exhaust gas temp			X	X	X
Power loss				X	
				— or —	
Percentage thrust				X	
				— or —	
Pressure ratio				X	

POWER INDICATORS FOR RECIPROCATING ENGINES

The power of the unsupercharged reciprocating engine is directly related to its speed, and so with the throttle at a corresponding setting, a tachometer system (described in Chapter 11) also serves as a power indicator,

With the supercharged engine, however, additional parameters come into the picture; fuel flow, increase of pressure ("boost" pressure) produced by the supercharger at the induction manifold, and on some engines either torquemeter pressure or brake mean effective pressure (bmep). Thus, power is monitored on four instruments. Fuel-flow indicators having already been described in Chapter 14; the following descriptions are confined to the remaining three instruments.

Manifold Pressure Gauges

Manifold pressure gauges, or "boost" gauges as they are more commonly termed, are of the direct-reading type and are calibrated to measure absolute pressure in pounds per square inch or in inches of mercury. Before considering a typical example, it will be useful to outline briefly the general principle involved in the supercharging of an aircraft engine.

The power output of an internal combustion engine depends on the density of the combustible mixture of fuel and air introduced into its cylinders at that part of the operating cycle known as the *induction stroke*. On this stroke, the piston moves down the cylinder, an inlet valve opens, and the fuel/air mixture, or charge prepared by the carburettor, enters the cylinder as a result of a pressure difference acting across it during the stroke. If, for example, an engine is running in atmospheric conditions corresponding to the standard sea-level pressure of $14\cdot7 \mathrm{lb/in^2}$, and the cylinder pressure is reduced to say, $2\mathrm{lb/in^2}$, then the pressure difference is $12\cdot7\mathrm{lb/in^2}$, and it is this pressure difference which "pushes" the charge into the cylinder.

An engine in which the charge is induced in this manner is said to be normally aspirated; its outstanding characteristic is that the power it develops steadily falls off with decrease of atmospheric pressure. This may be understood by considering a second example in which it is assumed that the engine is operating at an altitude of 10,000ft. At this altitude, the atmospheric pressure is reduced by an amount which is about a third of the sea-level value, and on each induction stroke the cylinder pressures will decrease in roughly the same proportion. We thus have a pressure of about $10\mathrm{lb/in^2}$ surrounding the engine and $1\frac{1}{2}\mathrm{lb/in^2}$ in each cylinder, leaving us with a little more than $8\cdot5\mathrm{lb/in^2}$ with which to "push in" the useful charge. This means then at 10,000ft only a third of the required charge gets into the cylinders, and since power is governed by the quantity of charge, we can only expect a third of the power developed at sea-level.

This limitation on the high-altitude performance of a normally-aspirated engine can be overcome by artificially increasing the available pressure so as to maintain as far as possible a sea-level value in the induction system. The process of increasing pressure and charge density is known as *supercharging* or *boosting*, and the device employed is, in effect, an elaborate form of centrifugal air pump fitted between the carburettor and cylinders and driven from the engine crankshaft through step-up gearing. It pumps by giving the air a very high velocity, which is gradually reduced as it passes through diffuser vanes and a volute, the reduction in speed giving the required increase in pressure.

Engine Power and Control Instruments 337

In order to measure the boost pressure delivered by the supercharger and so obtain an indication of engine power, it is necessary to have a gauge which indicates absolute pressure. The mechanism of a typical gauge working on this principle is illustrated in Fig 15.1. The measuring element is made up of two

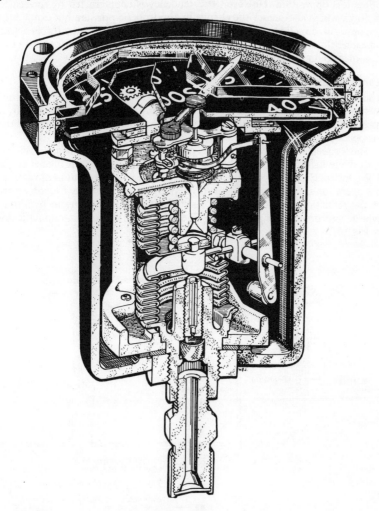

Fig 15.1 Typical manifold pressure gauge

bellows, one open to the induction manifold and the other evacuated and sealed. A controlling spring is fitted inside the sealed bellows and distension of both bellows is transmitted to the pointer via the usual lever, quadrant and pinion mechanism. A filter is located at the inlet to the open bellows, where there is also a restriction to smooth out any pressure surges.

When pressure is admitted to the open bellows it expands and causes the pointer to move over the scale and so indicate a change in pressure. With increasing altitude, there is a tendency for the bellows to expand a little too far because the decrease in atmospheric pressure acting on the outside of the bellows offers less opposition. However, this tendency is counteracted by the sealed bellows, which also senses the change in atmospheric pressure but expands in the opposite direction. Thus a condition is reached at which the forces acting on each bellows are equal, cancelling out the effects of atmospheric pressure so that manifold pressure is measured directly against the spring.

Torque Pressure Indicators

These indicators supplement the power indications obtainable from tachometers and manifold pressure gauges by measuring the pressures created by a torquemeter system, such pressures being interpreted as power available at the propeller shaft.

The torquemeter system forms part of the engine itself and is usually built-in with the reduction-gear assembly between the crankshaft and propeller shaft. The construction of a system depends on the type of engine, but in most cases the operation is based on the same principles; i.e. the tendency for some part of the reduction gear to rotate is resisted by pistons working in hydraulic cylinders secured to the gear casing. The principle is shown diagrammatically in Fig 15.2.

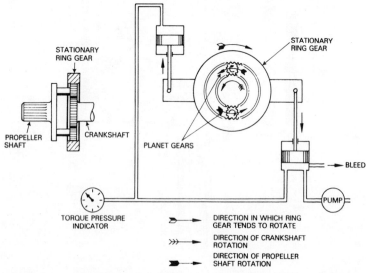

Fig 15.2   Principle of torquemeter

Oil from the engine system is supplied to the cylinders via a special torquemeter pump and absorbs the loads due to piston movement. The oil is thus subjected to pressures which are proportional to the applied loads or torques and are transmitted to the torque pressure-indicating system, which is normally of the remote-indicating synchronous type.

The brake horsepower is calculated by the following formula:

$$\mathrm{BHP} = \frac{pN}{K}$$

where p is the oil pressure, N the speed (rev/min) and K a torquemeter constant derived from the reduction gear ratio, length of torque arm, and number and area of pistons.

POWER INDICATORS FOR TURBOPROP ENGINES

Turboprop engines are, as far as power is concerned, similar to large supercharged piston engines; most of the propulsive force is produced by the propeller, only a very small part being derived from the jet thrust. They are therefore fitted with a torquemeter and pressure gauge system of which the oil pressure readings are an indication of the shaft horsepower. The torquemeter pressure gauge is used in conjunction with the tachometer and turbine gas-temperature indicators.

The indicating system used is governed by the particular type of engine, but there are two main systems in current use and their operating principles are based on the Desynn and alternating-current synchro methods of transmission.

Desynn Torque Pressure Indicating System

This system operates on the slab-Desynn principle described on page 236 and is used on Rolls–Royce Dart turboprop engines. In common with the other Desynn systems the transmitter, shown in Fig 15.3, comprises both mechanical and electrical elements.

The mechanical element consists of a Bourdon tube the open end of which is connected by a flexible hose to the supply line from the oil pump of the engine torquemeter. The free end of the Bourdon tube is connected to the brushes of the electrical element via a sector gear and pinion. A union mounted adjacent to the main pressure connection is connected to a capillary tube accommodated within the Bourdon tube and allows for the bleeding of the system. The transmitter is mounted in a special anti-vibration mounting, the whole assembly being secured to the engine itself.

With the engine running the pressure produced at the pistons of the torquemeter system is sensed by the transmitter Bourdon tube, the free end of which is distended so as to change the radius of the tube. The movement of the free end is magnified by the sector and pinion, which causes the brushes to be rotated over the slab-wound resistor. The resulting currents, and magnetic field produced in the indicator stator, position the rotor and pointer to indicate the torque pressure on a dial calibrated from 0 to 600 lb/in^2. During operation, and due to pulsations of torquemeter pressure, a certain amount of pointer fluctuation is possible, but this is limited to 30 lb/in^2 on either side of a mean torquemeter pressure reading.

Synchro Torque Pressure Indicating System

This system is an application of the alternating-current "power follow-up" synchro principle (see page 244).

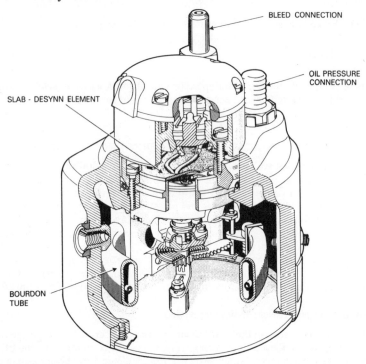

Fig 15.3 Desynn torque pressure transmitter

As may be seen from Fig 15.4, the mechanical element of a synchro torque pressure transmitter is very similar to that of the slab-Desynn type. The Bourdon tube, sector gear and pinion, however, are arranged to drive the rotor of a synchro. The transmitter is designed for mounting directly on to an engine and is connected by flexible tubing to the torquemeter system.

The indicator consists of a control transformer synchro connected to the transmitter synchro, a two-stage transistor amplifier, a two-phase servomotor, and two concentrically mounted pointers driven through a gearbox. The smaller pointer indicates hundreds of pounds and rotates in step with the synchro rotor, while the larger pointer rotates ten times as fast.

When the Bourdon tube senses a change in torquemeter system pressure it causes the transmitter synchro rotor to rotate and to induce a signal voltage in its stator which is then transmitted to the stator of the control-transformer synchro in the indicator. This signal results in a change in direction of the resultant magnetic field with respect to the control-transformer rotor position, thus inducing an error voltage signal in the rotor. The error signal is fed to the amplifier, which determines its direction, i.e. whether it results from an increase or a decrease in pressure, as well as amplifying it. The amplified signal is then applied to the control phase of the servomotor, which, via the gearbox, drives the pointers in the appropriate direction and also drives the rotor of the control transformer until it reaches a new null position at which no further error voltage signal is induced.

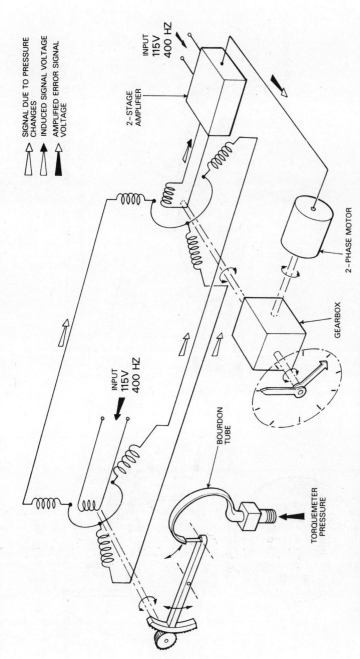

Fig 15.4 Principle of synchro-type torque pressure indicating system

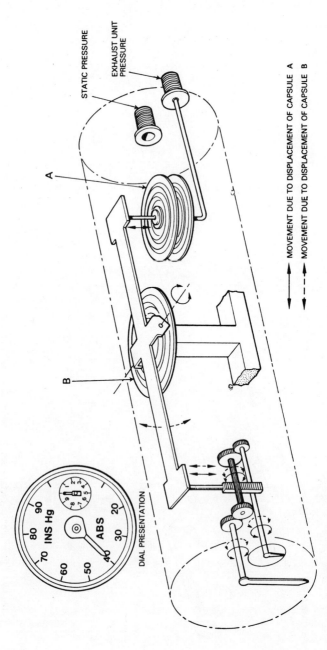

Fig 15.5 Power loss indicator

POWER INDICATORS FOR TURBOJET ENGINES

With turbojet engines the number of instruments required for power monitoring depends upon whether the engine employs a centrifugal or an axial type of compressor. The thrust of a centrifugal compressor engine is approximately proportional to the speed, so that the tachometer, together with the turbine gas-temperature indicator, may be used to indicate thrust at the specified throttle setting.

The thrust produced by an axial compressor engine does not vary in direct proportion to the speed, the thrust ratings being calculated in such a way that they must be corrected for variations in temperature and pressure prevailing at the compressor intake. Since compressor intake pressure is related to the outlet or discharge pressure at the turbine, then thrust is more accurately determined by measuring the ratio between these two pressures. This is done by using either a power-loss indicator, an engine pressure ratio indicator, or a percentage thrust indicator, in conjunction with the r.p.m., turbine gas-temperature and fuel-flow indicators.

Power Loss Indicating System

This system consists of a number of pressure-sensing probes mounted around the periphery of an exhaust unit, a static-pressure vent line and a sensitive absolute-pressure indicator.

A typical indicator, shown diagrammatically in Fig 15.5, comprises two capsule assemblies, one (A) open to exhaust unit pressure, and the other (B) evacuated and sealed and therefore acting as an aneroid and responding to the static pressure. Movement of the capsule assemblies in response to pressure changes is transmitted to two pointers via a beam and gear mechanism. The small pointer registers against a scale calibrated from 0 to 10 in Hg, while the larger pointer registers against a scale calibrated from 10 to 100 in Hg.

The exhaust unit pressure is admitted to the interior of capsule A while its exterior is subjected to the static pressure which is admitted to the indicator case. Its displacements are therefore a measure of the differential pressure equal to (exhaust unit pressure − static pressure). As capsule A expands or contracts it rocks the beam about its pivoting point on capsule B and establishes a certain angular position corresponding to the differential pressure across capsule A. At the same time capsule B responds to static pressure changes so that its expansion or contraction will move the beam to what may be termed its "differential pressure" position. As the beam is coupled to the gear mechanism of the pointers, these will be finally positioned over their respective dials to indicate the total exhaust unit pressure proportional to the algebraic sum (exhaust unit pressure − static pressure) + static pressure.

Percentage-thrust Indicator

A percentage-thrust indicator operates on the same principle of pressure sensing as a power-loss indicator, but as will be noted from Fig 15.6, it has the advantage of indicating percentage of thrust in quantitative form, over the range 50 to 100%. In addition, it incorporates a manually controlled device permitting the thrust indications to be compensated for variation in ambient atmospheric conditions. The compensation is accomplished by rotating a setting knob, which adjusts a digital counter (in some instruments a scale may

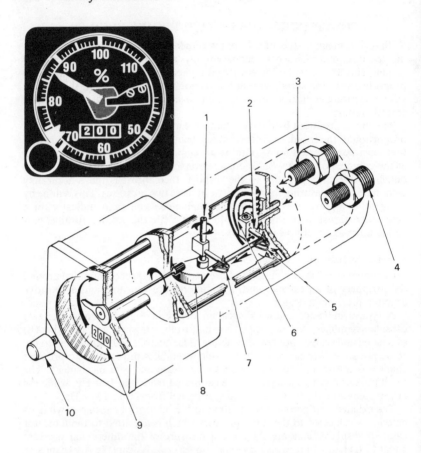

Fig 15.6 Percentage-thrust indicator

1. Rocking shaft
2. Calibrating arm
3. Exhaust unit pressure connection
4. Static pressure connection
5. Overload spring
6. Capsule
7. Overload spring
8. Sector gear
9. Digital counter (atmospheric datum)
10. Counter setting knob

be fitted) and at the same time rotates the complete mechanism and positions the pointer to a new datum value on the main dial.

With this compensation applied, the instrument normally indicates 100% thrust as a minimum take-off value under conditions least favourable to engine performance. Under more favourable conditions the engine performance may indicate a take-off value greater than 100% thrust.

The counter is of the three-digit display, and each number set on it corresponds to an appropriate ambient atmospheric condition obtained from performance curves plotted for specific aircraft/engine combinations.

TURBINE TEMPERATURE CONTROL

The power developed by a turbine engine is dependent on two main factors: the air mass flow through it and the temperature drop. The air mass flow varies with engine speed and also with air density, which in turn is determined by altitude, atmospheric temperature and forward speed. Temperature drop is the difference between the temperatures immediately before and after the turbine and is therefore a measure of the energy extracted by the turbine.

It is thus apparent that the temperature drop will be a maximum and so indicate the maximum energy extraction if the gas temperature at the entry to the turbine is maintained at the highest level. There is, however, a practical limitation to this temperature brought about by the effects on the material of the turbine blades and consequently on their life. For this reason, optimum temperatures are established at which the maximum power may be obtained without impairing the structural integrity of the turbine blading, and the operating conditions are carefully controlled to ensure that such limitations are not exceeded.

Control of the conditions can be instituted at ground level, but in flight a turbine must operate under changing conditions of atmospheric temperature, density and forward speed, and as already mentioned, these variables determine the air mass flow through the turbine. For given atmospheric conditions, the air mass flow is controlled by the engine speed, and to maintain the maximum turbine entry temperature appropriate to these conditions, the flow of fuel to the engine must be controlled so as to match the air mass flow.

This process of fuel flow control is generally known as *fuel trimming*; depending on particular engine installations it may be effected by an electro-mechanical system under the direct control of the pilot, or by a temperature control system which automatically monitors the fuel flow in response to signals from the thermocouples of the standard exhaust-gas temperature-measuring system.

Fuel Trim Indicating System

An example of the electro-mechanical method is the one adopted for the trimming of the fuel flow to the Rolls–Royce Dart turboprop engine and serves as a useful illustration.

The air mass and fuel flows are matched to the designed ratio, in the first instance, by mechanically interconnecting the fuel throttle valve with the r.p.m. controls. This results in an optimum gas temperature for any selected engine speed.

To compensate for the changes in air mass flow, an electric actuator operates a differential compound lever mechanism incorporated as part of the interconnection between throttle valve and r.p.m. controls. Electrical power to the actuator is controlled by a switch which is accessible to the pilot and is placarded INCREASE and DECREASE, indicating the trimming condition of fuel flow required and consequently the change of turbine temperature. The pilot must, of course, have some means of knowing by how much the fuel flow should be trimmed, and so a fuel-trim position indicating system is provided for his use in conjunction with datum position tables or a datum computer supplied by the engine manufacturer.

The indicating system consists of a position transmitter and indicator

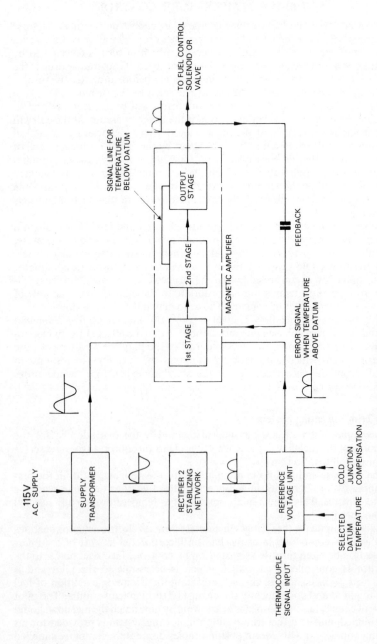

Fig 15.7 Typical automatic temperature-control amplifier

operating on the basic Desynn principle described in Chapter 10 (page 230). The transmitter is mechanically connected to the trim actuator by a linkage suited to the engine installation and electrically connected to the indicator, which is usually mounted on the control pedestal in the cockpit. The scale of the indicator is graduated from 0 (full decrease) to 10 (full increase), corresponding to the range 0 to 100% fuel trim.

The percentage increase or decrease is related to the prevailing air temperature and the pressure altitude, and is obtained from the datum position tables or computer. When the required value has been selected, the actuator is switched on so that its shaft will retract or extend, depending on whether an increase or decrease of the fuel flow is required. Movement of the actuator shaft positions the throttle-valve lever, via control rods and the differential compound lever which permits trimming of the fuel without disturbing the setting of the r.p.m. controls. The actuator shaft also positions the brushes over the toroidal resistance of the transmitter, the voltage combinations so produced being supplied to the indicator stator windings. Since the indicator rotor is a permanent magnet, it aligns itself with the resultant magnetic field induced in the stator, and at the same time moves the pointer to indicate the change in fuel trim.

Automatic Temperature Control System

A logical development of the fuel trimming process, particularly since it is required to maintain turbine temperatures within specified limits, is to utilize the signals generated by the thermocouples of the standard temperature indicating system and so let these do the work of automatically changing the fuel flow. Experiments along these lines resulted in a specially designed amplifier unit which, on being connected to the thermocouples, amplifies the signals produced above a preselected datum temperature and supplies them to a solenoid-operated servo-valve which then progressively reduces the fuel flow to restore the datum temperature condition.

A block functional diagram of such an amplifier is shown in Fig 15.7 and may serve as a basis for the explanation of temperature control systems in current use.

The thermo-e.m.f. is fed as an input signal to a terminal block on the amplifier, the terminals forming the cold junction of the control system. Compensation for temperature changes at this junction is effected by a bridge circuit, one arm of which changes its resistance with changes of temperature. In addition to the thermocouples, the bridge circuit is also connected to a reference-voltage unit, the purpose of which is to inject a voltage in opposition to that of the thermocouples and corresponding with the desired operating temperature range of the engine. This voltage is stabilized against changes in supply voltage and frequency (115V, 400Hz) and ambient temperature, and is selected by a temperature selector.

The amplifying section is in three main stages: amplifier stages 1 and 2, and an output stage connected to the solenoid valve. All three stages are of the magnetic amplifier type.

When the engine is running at the selected temperature, the e.m.f. from the thermocouple is opposed by an equal voltage from the reference-voltage unit. so that there is no input signal to stage 1 of the amplifier.

If the jet pipe temperature is below the selected value, the reference voltage is greater than that of the thermocouples and this provides a reverse-polarity input to stage 1. This reverse signal is, however, blocked by a network connected between stage 2 and the output stage so that no output current is obtained.

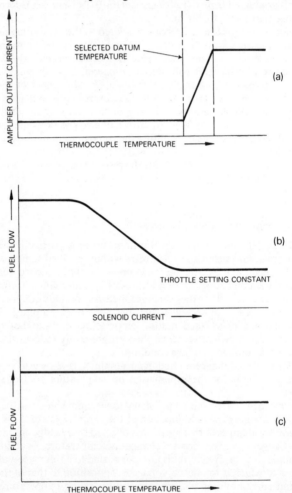

Fig 15.8 Operating characteristics of a typical temperature control system
(a) Amplifier
(b) Fuel control system
(c) Combined control characteristic

When the jet pipe temperature rises above the selected value, and this is the more critical situation, the predominating voltage is that from the thermocouples. As this voltage is of the correct polarity the resulting signal is fed to stages 1 and 2 and, after amplification, to the solenoid valve via the output stage. The valve operates so as to restrict the fuel supply to the engine, thus reducing the jet pipe temperature and thermocouple e.m.f. until it balances the reference voltage once again. The solenoid valve is then released and the normal fuel flow is restored.

The opeiating characteristics of a typical temperature control system are shown graphically in Fig 15.8. It will be noted from diagram (*a*) that, when the temperature reaches the limiting value, the amplifier output continues to increase, reaching a constant maximum value after a small temperature increase of approximately 8°C. The solenoid-current/fuel-flow characteristic depends on the type of fuel system employed, but it is usually of the form shown at (*b*), which clearly indicates the decrease in the fuel flow with the increase in solenoid current. A combination of (*a*) and (*b*) gives us the final control characteristic shown at (*c*).

"Hunting" of the solenoid valve which would give rise to a fluctuating fuel flow and engine thrust, is prevented by feeding back some of the output current into the amplifier stages, thus increasing the time lag.

COMBINED TEMPERATURE AND R.P.M. CONTROL SYSTEM

This system is a further development in the field of automatic control of engine power, and is one in which fuel flow is regulated by combining the signals from thermocouples with those supplied by a tachometer generator. A block diagram of the system is shown in Fig 15.9.

The temperature control channel operates on the same principle as the system described in the preceding paragraphs, and so we need only consider the r.p.m. control channel.

The three-phase output from the tachometer generator is supplied to an r.p.m. discriminator section which is made up of two elements, the purpose of which is to provide two independent voltages from the generator input. One element is of the resistance type so that its voltage is largely independent of the generator frequency, and the other element is of the reactance type, i.e. it contains capacitors and so its voltage is proportional to frequency and therefore to engine speed. Both these voltages are then rectified and applied in opposition to the r.p.m. control channel amplifier, which is of the two-stage magnetic type. The voltages may be equalized at any desired engine speed by adjustment of the resistance-type element with the aid of a selector. The output of the r.p.m. control channel amplifier is in turn fed to the output stage, which, as may be seen from the diagram, is common to both r.p.m. and temperature control channels. At the selected datum speed, the current from the output stage is the standing current in the fuel-valve solenoid.

When the engine speed is below the selected datum, the r.p.m. control channel signal is suppressed by a discriminator network similar to that used for the temperature channel. If, however, the engine speed rises above datum, the voltage output from the reactive element of the r.p.m. discriminator section will predominate and pass a signal on to the solenoid valve, which reduces the fuel flow to restore the datum speed condition.

The circuitry between both control channels is so arranged that the solenoid valve remains under the control of the r.p.m. channel until such time as the jet pipe temperature exceeds the datum limit, when the temperature control channel with its greater amplification overrides the speed signal and assumes control of the fuel flow to reduce turbine temperature.

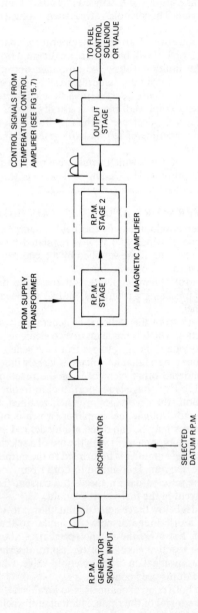

Fig 15.9 Typical temperature and r.p.m. control system

The application of this system depends upon the requirements of a particular engine; for example, speed and temperature limiting may be required at a fixed datum ("top limiting"), or it may be required to be variable ("range control"). In either case, appropriate datum selector units are used with the amplifiers.

An example of "top limiting" is illustrated graphically in Fig 15.10 and relates to an aircraft climbing from ground level. Under these conditions,

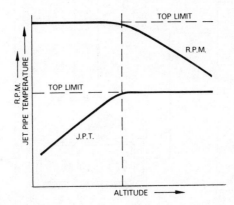

Fig 15.10 Top limiting

engine speed is the limiting factor and maximum take-off power is required. The solenoid valve modifies the fuel flow to maintain constant r.p.m. until the jet pipe temperature rises to its limiting value. When this limit is reached, the temperature channel takes full control and overrides the governing action of the r.p.m. channel, thus reducing the fuel flow to avoid exceeding the datum temperature.

ENGINE VIBRATION MONITORING AND INDICATING SYSTEMS

Engine vibration is, of course, something which is unwanted, but unfortunately it cannot be entirely eliminated even with turbine engines, which have no reciprocating parts. It can only be kept down to the lowest possible level.

During operation, however, there is always the possibility of vibration occurring in excess of acceptable levels, as a result of certain mechanical troubles. For example, a turbine blade may crack or "creep" or an uneven temperature distribution around turbine blades and rotor discs may be set up: either of these troubles will give rise to an unbalanced condition of the main rotating assemblies. In order, therefore, to monitor vibration and to indicate when the maximum amplitude on any engine exceeds a preset level, systems have been developed which come within the engine control group of instrumentation.

A system consists essentially of a vibration pick-up unit mounted on the engine at right angles to its axis, an amplifier monitoring unit and a moving-coil microammeter calibrated to show vibration amplitude in thousands of an inch (mils). A block diagram of the system is shown in Fig 15.11.

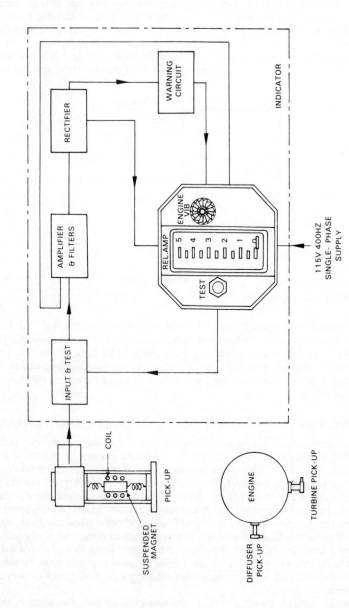

Fig 15.11 Turbine vibration monitoring system

The pick-up unit is a linear-velocity detector that converts the mechanical energy of vibration into an electrical signal of proportional magnitudes. It does this by means of a spring-supported permanent magnet suspended in a coil attached to the interior of the case.

As the engine vibrates, the pick-up unit and coil move with it; the magnet, however, tends to remain fixed in space because of inertia. The motion of the coil causes the turns to cut the field of the magnet thus inducing a voltage in the coil and providing a signal to the amplifier unit. The signal, after amplification and integration by an electrical filter network, is fed to the indicator via a rectifying section.

An amber indicator light also forms part of the system, together with a test switch. The light is supplied with direct current from the amplifier rectifying section and it comes on when the maximum amplitude of vibration exceeds the preset value. The test switch permits functional checking of the system's electrical circuit.

In some engine installations, two pick-up units may be fitted to an engine, one monitoring vibration levels around the turbine section and the other around the diffuser section. In this case, a two-position switch is included in the monitoring system so that the vibration level at each pick-up may be selected as required and read on a common indicator.

QUESTIONS

15.1 Briefly explain the principle of engine supercharging.

15.2 Describe the construction and explain the operation of the instrument used for measuring supercharger delivery pressure.

15.3 What would be the indications of gauges calibrated in pounds per square inch and in inches of mercury, under static conditions of an engine at standard atmospheric pressure?

15.4 What is the function of a torque pressure-indicating system? Describe the construction and operation of a system with which you are familiar.

15.5 List the instruments required for monitoring the power of turbojet engines.

15.6 Sketch a typical engine power-loss indicating system and explain its principle of operation. (SLAET)

15.7 Describe a method of utilizing the voltage generated by exhaust-gas thermocouples for controlling the fuel flow and gas temperature.

15.8 Describe the operation of an engine vibration indicator. Show with the aid of a sketch the construction of the sensing device. (SLAET)

15.9 List the complete engine and power plant instrumentation which you would expect to find in a multi-engined aircraft. (SLAET)

16 Integrated Instrument and Flight Director Systems

The need for integrating the functions and indications of certain flight and navigation instruments resulted in the main from the increasing number of specialized radio aids linking aircraft with ground stations. These were developed to meet the demands of safe en-route navigation and to cope with increasing traffic congestion in the air space around the world's major airports.

The required information is processed by a multiplicity of "black boxes" which can be stowed in electrical compartments and radio racks, but in order that the necessary precision flying may be executed, information must still be presented to the pilot. This requires more instruments and more instruments could mean more panel space. The method of easing the problem was to combine related instruments in the same case and to compound their indications so that a large proportion of intermediate mental processing on the part of the pilot could be bypassed and the indications more easily assimilated.

In some respects, integration of instruments is not new; for example, a combined pressure and temperature indicator was in use long before the present state of the art. Another early example and one of those around which most of today's fully integrated flight and navigation instrument systems have been built up is the *Radio Magnetic Indicator* (RMI). This instrument combines and presents information from three separate sources on one dial; magnetic heading from a remote-indicating compass, the bearing from a *very-high-frequency Omnidirectional Range* (VOR) ground station, and the bearing from an *Automatic Direction Finding* (ADF) station.

During that phase of a flight involving the approach to an airport runway, it is essential for a pilot to know, among other things, that he is maintaining the correct approach attitude. Such information can be obtained from the gyro horizon and from a special ILS meter which responds to vertical and horizontal beam signals radiated by the transmitters of an *Instrument Landing System* located at the airport. It was therefore a logical step in the development of integrated instrument systems to include the information from both these instruments.

Figure 16.1 illustrates the dial presentations of two instruments which, although made by different manufacturers and consequently not used in combination, serve as useful examples of how the information from the various sources referred to is integrated in practice.

The director horizon has the appearance of a conventional gyro horizon, but unlike this instrument the pitch and bank indicating elements are electrically controlled from a remotely located vertical gyro unit. Furthermore, it employs a different method of referencing the elements. These features are common to all integrated flight systems. The horizon bar and bank scale are marked on the background disc which is monitored by bank signals from the vertical gyroscope and rotates about the fore-and-aft axis. The miniature aircraft which, in the

Integrated Instrument and Flight Director Systems 355

instrument illustrated is called the *pitch bar*, is monitored by pitch signals and moves in a vertical plane above and below the centre-line of the instrument.

The approach attitude of an aircraft with respect to ILS vertical-beam (glide-path) signals and horizontal-beam (localizer) signals is indicated by independent pointers monitored by the relevant ILS receiver channels. The glide slope pointer is referenced against a vertical scale and the pitch bar, and shows the displacement of the aircraft above or below the glide path. Displacement of the aircraft to the left or right of the localizer beam is indicated by angular deflections of the localizer or steering pointer. ILS information is selected by turning a switch at the front of the instrument (see Fig 16.1) to the appropriately marked position.

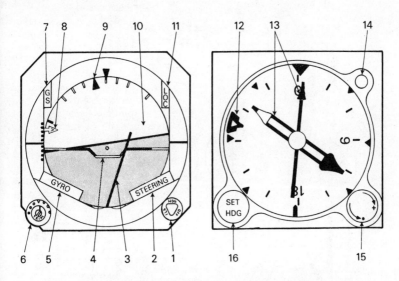

Fig 16.1 Dial presentations of typical flight director system instruments

1. Mode selector knob (HDG/ILS)
2. Steering flag
3. Steering pointer
4. Pitch bar
5. Gyro flag
6. Pitch trim knob
7. Glide slope flag
8. Glide slope pointer
9. Bank pointer
10. Horizon disc and bar
11. Localizer flag
12. Set heading cursor
13. RMI pointers (ADF/VOR)
14. Annunciator window
15. Synchronizing knob
16. Set heading knob

Electrical interconnection of the director horizon components primarily concerned with pitch and bank attitude information is shown in Fig 16.2. Whenever a change of aircraft attitude occurs, signals flow from pitch and bank synchros disposed about the relevant axes of the vertical gyroscope to the corresponding synchros within the instrument. Error signals are therefore induced in the rotors and after amplification are fed to the servomotors, which rotate to position the pitch bar and horizon disc to indicate the changing attitude of the aircraft. At the same time, the servomotors drive the synchro rotors to the "null" position.

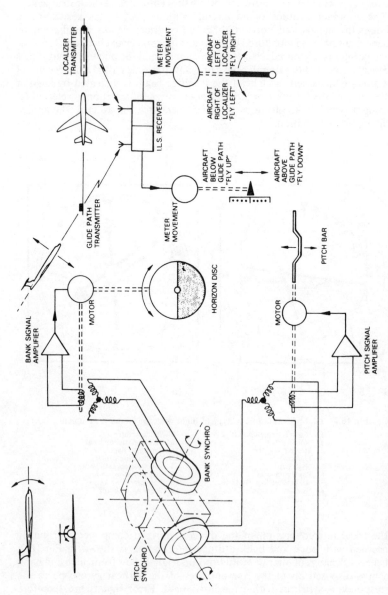

Fig 16.2 Electrical interconnection of director horizon elements

Figure 16.2 also shows the interconnection of the glide slope and localizer pointer with the ILS. During an ILS approach the receiver on board the aircraft detects the signals beamed from ground transmitters in vertical and horizontal planes. If the aircraft is above the glide path, signals are fed to the meter controlling the glide slope pointer causing it to be deflected downwards against the scale, thus directing the pilot to bring the aircraft down on to the glide path. An upward deflection of the pointer indicates flight below the glide path and therefore directs that the aircraft be brought up to the glide path. The pointer is also referenced against the pitch bar to indicate any pitch correction required to capture and hold the glide path. When this has been accomplished, the glide slope pointer and pitch bar are matched at the horizontal centre position.

If, during the approach, the aircraft is to the left of the localizer beam and runway centre-line, the localizer pointer is deflected to the right directing that the aircraft be banked to the right. Flight to the right of the localizer beam causes pointer deflection to the left, directing that the aircraft be banked to the left. When either of these directions has been satisfied, the pointer is positioned vertically through the centre position of the horizon disc.

When the HDG-ILS switch (see Fig 16.1) is turned to the HDG position the localizer pointer is actuated by combined heading and bank signals. The purpose of the TRIM knob is to alter the position of the pitch bar during HDG operation, to allow the use of a pitch reference, other than horizontal, to be used. Rotation of the knob operates a gear train to move the pitch bar down for clockwise rotation, and upward for anticlockwise rotation. The adjustment range is $+20°$ to $-15°$ in $5°$ increments marked on the knob.

The four warning flags come into view when all circuits associated with the instrument are isolated from electrical power, and also if signals are weak or circuits do not function properly during complete system operation.

The course indicator shown in Fig 16.1 is a special version of RMI, the function of which was mentioned on page 354, and is in fact a component part of a remote-indicating compass system employing a master gyro unit. It consists of three synchros: one controlling the heading card and monitored by the gyro heading synchro, the second controlling a double-bar pointer, and the third controlling a single-bar pointer. The pointer synchros are monitored by radio signals from VOR and ADF transmitters to indicate against the compass card the bearing of each transmitter. Each pointer has an arrow-shaped head so that when rotating they also indicate whether the aircraft is being flown towards or away from a transmitter.

The electrical interconnection of the components is shown in Fig 16.3, and an example of the indications presented by the instrument is given in Fig 16.4, which also shows how the position of an aircraft can be fixed from the instrument indications. The positions of the transmitter are shown on navigational charts, so that if lines are drawn from the transmitters as reciprocals of the bearings relative to magnetic North, their points of intersection mark the position of the aircraft at the time the readings were taken.

Another version of course indicator is illustrated in Fig 16.5; this is the one used in conjunction with the director horizon of Fig 16.1. It combines magnetic heading and radio bearing indications to form a plan view or map-like display of the aircraft with respect to a selected course indicated by the miniature aircraft symbol fixed at the centre of the instrument. The azimuth ring or compass card is monitored by signals from a master directional gyro unit and

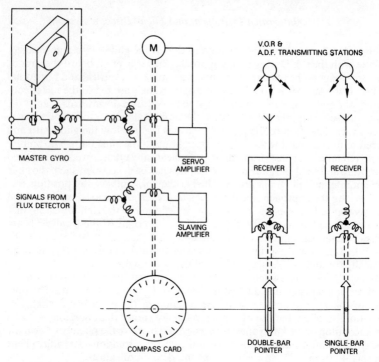

Fig 16.3 Course indicator components

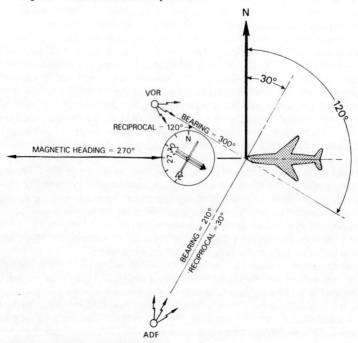

Fig 16.4 Examples of indications

indicates magnetic heading against the lubber line. A compass warning flag comes into view when the signal circuits from the compass are not functioning properly. The interconnection of the course indicator element is shown in Fig 16.6. Located centrally within the azimuth ring is a course deviation element comprising a course bar, course arrow, a to-from arrow, and a course deviation scale. The element rotates with the azimuth ring as it responds to changes of heading signal.

The course bar is controlled by a meter movement and registers against the 5-dot deviation scale the amount of aircraft deviation left or right of an ILS localizer beam or a selected VOR transmitter bearing. On an ILS localizer frequency each dot represents a $\frac{1}{2}°$ deviation left or right, and on a VOR bearing a

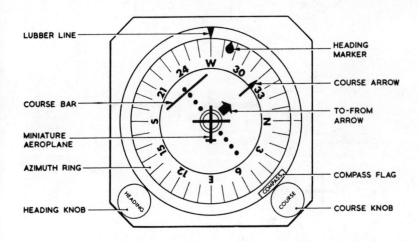

Fig 16.5 Course indicator

2° deviation. When the aircraft is on course the bar moves to the centre and is aligned with the head and tail of the course arrow. The course arrow shows the direction of the localizer or the VOR transmitter and can be independently set by the COURSE knob. The to-from arrow is controlled by a meter movement and indicates flight of the aircraft towards or away from the VOR transmitter. During the ILS mode of operation the arrow is withdrawn from view.

A selected magnetic heading may be set in by rotating the HEADING knob to move the heading marked round the azimuth ring. The knob also changes the position of a selected heading synchro rotor to produce an error signal which is fed to the steering pointer meter in the director horizon (see Fig 16.1), and also to the appropriate circuit of an automatic control system. Once set, the heading marker rotates with the azimuth ring, giving a continuous display of the selected heading and any actual heading deviation.

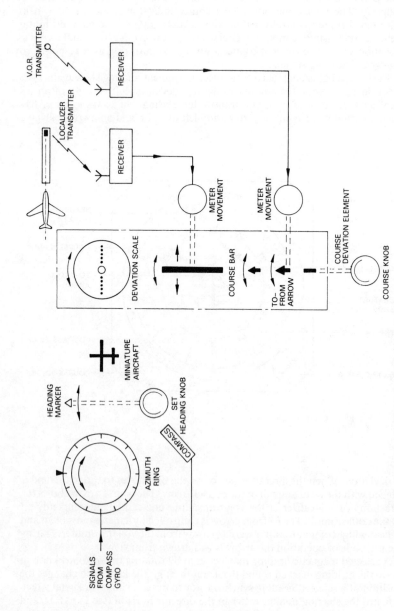

Fig 16.6 Interconnection of course indicator elements

QUESTIONS

16.1 Draw a diagram to illustrate the display of a typical director horizon.

16.2 What is the function of the warning flags provided in director horizons and course indicators and when do they come into view?

16.3 With the aid of a diagram explain how the position of an aircraft can be obtained from the data provided by a course indicator.

16.4 With the aid of a block diagram describe in outline a typical integrated flight system. (SLAET)

Principal Symbols and Abbreviations

Symbols for quantities are in *italic* type, and abbreviations for the names of units (unit symbols) are in ordinary type.

A	ampere	mV	millivolt
a	speed of sound	m.p.h.	mile per hour
B	magnetic flux density	mm H$_2$O	millimetre of water
bhp	brake horsepower	N	newton
C	capacitance	N	number of turns of a coil
F	farad	p	pressure
F	force	pF	picofarad
ft/min	foot per minute	R	resistance
ft/h	foot per hour	r.p.m. ⎫	
ft/s^2	foot per second per second	rev/min ⎭	revolution per minute
g	acceleration due to gravity	T	period, periodic time
H	magnetic field strength	V	volt
Hz	hertz (cycle per second)	V	velocity
I	electric current		voltage
	moment of inertia	W	weight
in Hg	inch of mercury	Wb	weber
K	relative permittivity (alternative to ϵ)	α	temperature coefficient of resistance
lb/gal	pound per gallon	μF	microfarad
lb/in^2	pound per square inch	ρ	density
M	Mach number		resistivity
	torque	Φ	magnetic flux
m	magnetic moment	Ω	ohm
	mass	ω	angular velocity (radians per second)
mA	milliampere		
mb	millibar		

Solutions to Numerical Questions

8.6 Coefficient $B = \dfrac{\text{Deviation on East} - \text{Deviation on West}}{2}$

$\qquad\qquad = \dfrac{+4 - (-2)}{2} = \dfrac{+2}{2} = +1°$

Coefficient $C = \dfrac{\text{Deviation on North} - \text{Deviation on South}}{2}$

$\qquad\qquad = \dfrac{+4 - (-1)}{2} = \dfrac{+3}{2} = +1.5°$

Coefficient $A = \dfrac{\text{Deviation on N} + \text{NE} + \text{E} + \text{SE} + \text{S} + \text{SW} + \text{W} + \text{NW}}{8}$

$\qquad\qquad = \dfrac{(+4) + (+2) + (+4) + (+3) + (-1) + (-2) + (-2) + (0)}{8}$

$\qquad\qquad = \dfrac{+8}{8} = +1°$

12.4(*b*) $40°C = 40 + 273.15 = 313.15\,K$

$\qquad\qquad = \dfrac{40 \times 9}{5} + 32 = 104°F$

12.5 $77°F = (77 - 32) \times \dfrac{5}{9} = 25°C$

12.7(*a*) $R_T = R_1 + R_2 + R_3 = 20 + 15 + 40 = 75\,\Omega$

12.7(*b*) $\dfrac{1}{R_T} = \dfrac{1}{R_1} + \dfrac{1}{R_2} + \dfrac{1}{R_3} = \dfrac{1}{8} + \dfrac{1}{12} + \dfrac{1}{7} = 0.125 + 0.083 + 0.143 = 0.351$

so that

$R_T = \dfrac{1}{0.351} = 2.847\,\Omega$

The applied voltage is 24 V; therefore, from the expression $I = V/R_T$, the current flowing in the series circuit is 0·32 A, and in the parallel circuit, 8·424 A.

12.8 For a parallel circuit containing only two resistances,

$R_T = \dfrac{R_1 R_2}{R_1 + R_2} = \dfrac{20 \times 45}{30 + 45} = 13.84\,\Omega$

12.9(*b*) Cross-sectional area of wire is $a = \dfrac{\pi}{4} \times 0.08^2 = 0.00503\,\text{cm}^2$

The resistance is $R = \rho\,\dfrac{l}{a} = 1.7 \times 10^{-6} \times \dfrac{1200}{0.00503} = 0.406\,\Omega$

Index

ABSOLUTE ALTITUDE, 77
Absolute permittivity, 314
Absolute zero, 267
Acceleration errors
   compasses, 166–168
   gyro horizons, 135
Accelerometer, 223
Aclinic line *see* Magnetic equator
A.C. ratiometer *see* Ratiometer pressure gauge
A.D.F., 354, 357
Agonic lines, 156
Air data computer, 103
Air data systems, 103
Aircraft magnetic components
   hard-iron, 203
   rod components, 208
   soft-iron, 206
   total effects, 209
Airspeed indicators, 85
Airspeed switch units, 95
Alpha-numeric display, 29
Alternate pressure sources, 63
Altimeter, 72
Altitude
   absolute, 77
   errors, 76
   switches, 81
   warning systems, 81
Ampere per metre, 152
Aneroid barometer, 72
Angle of dip *see* Magnetic dip
Angular momentum, 111
Annual change, 156
Annunciator *see* Synchronizing indicators
Anti-backlash gear, 18
Aperiodic compass *see* Direct reading compasses
Apparent A, 214
Apparent tilt, 117
Apparent veer, 117
Artificial horizon *see* Gyro horizon
Atmospheric pressure, 69
Atmospheric temperature, 69

BACKLASH, 17
B/H curve, 182

Balancing coil, 276
"Banana-slot" compensator, 89
Bank indication
   ball-in-tube method, 144, 146
   gravity-weight method, 144, 145
Barometric pressure setting, 73, 77
Base metal thermocouple *see* Thermocouple materials
"Basic six" layout, 43
"Basic T" layout, 43
Bellows *see* Pressure measurement
Bimetal strip, 19 *see also* Cold-junction compensator
Boiling point, 266
Boost gauges *see* Manifold pressure gauges
Blind flying conditions, 2
Blind flying panel, 3, 43
"Bottom heaviness" *see* Pendulosity errors
Bourdon tube, 10, 299, 301, 305
Bridge lights *see* Instrument illumination

CABIN ALTIMETERS, 81
Caging knob, 172
Capacitance
   governing factors, 314
   in a.c. circuits, 315
   in series and parallel, 315
   principle, 313
   units of, 314
Capacitance-type fuel quantity indicators, 312
Capacitive index, 320
Capacitive reactance, 316
Capillary pressure gauge, 301
Capsules *see* Pressure measurement
Card compass, 160, 162 *see also* Direct reading compasses
Ceramic type metering unit *see* Metering units
Characterized tank units, 324
Circular scale, 23
Coefficients of deviation *see* Deviation coefficients
Coercive force, 183
Coercivity, 183

366 Aircraft Instruments

Cold-junction compensator, 290
Compass operating modes, 199
Compass swinging, 215
Compass system monitoring, 190
Compensated gauge system, 321
Compensating leads, 293
Compensation for dip, 162
Compensator tank unit, 321, 323
Component P, 204
Component Q, 204
Component R, 204
Compressibility error, 85
Conduction, 266
Control transformer, 244, 340
Controlling system, 279, 283
Convection, 266
Coloured displays, 31
Coupling element, 10
Course indicator, 357
Critical Mach number, 91
Cross-coil ratiometer, 281
Cyclic error see Desynn system
Cylinder-head temperature indicator, 287

DATA SYNCHRO, 244
Dead beat indication, 280
Dead weight tester see Pressure measurement
Declination see Magnetic variation
"Demand and response" data, 37
Density altitude, 77
Desynn system
 basic, 230
 cyclic error, 233
 micro, 235
 slab, 236
Detecting element (instrument), 10
Deviation, 203
Deviation coefficients, 212
Deviation compensation devices, 165, 166, 216
Diaphragms see Pressure measurement
Dielectric constant, 314
Digital display, 29
Directional gyroscope, 170–177
Directional gyroscope elements, 187
Directional gyroscope monitoring, 190
Director displays, 35
Director horizon, 354
Direct-reading compasses
 acceleration errors, 166–168
 aperiodic, 158
 dip compensation, 162, 164
 functions, 150

Direct-reading compasses—(contd.)
 heading presentation, 160
 liquid damping, 158
 liquid expansion compensation, 159, 164
 magnet systems, 158
 turning errors, 168–170
 typical compasses, 166–168
Direct-reading pressure gauge, 296, 301
Displays
 high-range long-scale, 25
 qualitative, 23
 quantitative, 23
Diurnal change, 156
Double-tangent mechanism, 16
Drag elements see Tachometers
Drains, 67
Drift, 172, 173
Dual-indicator display, 31
Duplicate instruments, 9
Dynamic counter display, 31

EARTH'S ATMOSPHERE, 68
Earth gyroscope, 117
Earth's magnetic components, 157, 207
Earth-rate error, 172, 173, 201
Elastic pressure-sensing elements, 299
Electric gyro horizon, 123
Electric zero, 242
Electrical temperature indicators
 resistance, 277
 thermoelectric, 277, 284–294
Electromagnetic compensator see Deviation compensation devices
Electromagnetic induction, 239
"Empty" adjustments, 325
Engine speed indicators see Tachometers
Engine vibration indicator, 351
Erection cut-out see Erection error compensation
Erection devices, directional gyroscopes, 172
Erection errors, 135
Erection error compensation
 erection cut-out, 138
 inclined spin axis, 137,'141
 pitch-bank erection, 139
Erection rate, 134
Erection systems, gyro horizon
 ball type, 125
 fast-erection systems, 131
 pendulous vane unit, 125
 torque motor and levelling switch, 127
Eureka spool, 294

Exhaust gas temperature indicators, 284, 287, 289
Exosphere, 69
Extension leads, 293
External circuit resistance, 293

FARAD, 314
Fast-erection systems
  electromagnetic method, 133
  fast-erection switch, 133, 139
  purpose, 131
Fatigue meter, 226
Fixed points, 266
Flexible drive shafts, 252
Flight director systems, 354
Float-type fuel-quantity indicators, 312
Flow lines, 47
Fluorescent dial markings *see* Instrument illumination
Flux detector elements, 178
Flux lines, 150
Free gyro mode *see* Compass operating modes
Free gyroscope, 111, 117
Fuel flowmeters, 326
Fuel quantity measurement, 312
Fuel quantity by weight, 320
Fuel trim indicator, 345
Fuel trimming, 345
"Full" adjustments, 325
Fundamental interval, 266 *see also* Temperature measurement

GEARS, 17
Geographic poles, 155
Gimbal lock, 121
Gimbal ring balancing, 173
Gimbal system, 111
Gimballing effects, 135
Gimballing errors, 174
Glide slope, 37, 355, 357
Grid ring, 166
Grid steering, 160, 165
Gyro horizon
  electric, 123
  erection systems, 125
  presentations, 35, 120
  principle, 119
  vacuum-driven, 122
Gyroscope, 111
Gyroscope levelling, 195
Gyroscopic rigidity, 111, 113, 115

HAIRSPRINGS, 18
Hard iron, 154 *see also* Hard iron magnetism

Hard iron magnetism, 203 *see also* Aircraft magnetic components
Head-up displays, 39
Heat, 265
High-range long-scale display, 25
Hysteresis curve, 182

ICAO STANDARD ATMOSPHERE, 70
Ice point, 266
ILS indicator, 36 *see also* Integrated instrument systems
Immersion thermocouple, 287 *see also* Thermocouple types
Inclined spin axis *see* Erection error compensation
Indicated altitude, 75
Indicating element, 10
Induction motor, 123
Inductor-type pressure gauge *see* Ratiometer pressure gauges
Inferred-density system, 321
Instrument displays, 23
Instrument elements, 10
Instrument grouping
  flight instruments, 43
  power plant instruments, 45
Instrument illumination
  bridge lights, 51
  fluorescent dial markings, 50
  pillar lights, 51
  Plasteck system, 53
  ultra-violet flood lighting, 50
  wedge-type lighting, 52
Instrument landing system, 36, 354, 357 *see also* ILS indicator
Instrument layouts *see* Instrument grouping
Instrument mechanism, 10
Instrument panels, 8, 39
Integrated director display *see also* Flight director systems, 37
Integrated fuel flowmeter, 329
Integrated instrument systems, 354
Invar, 19
Ionosphere, 69
Isoclinals, 157
Isodynamic lines, 158
Isogonal lines, 156

KELVIN SCALE *see* Temperature measurement
Knot, 86

LAPSE RATE, 69
Latitude control, 201

Aircraft Instruments

Levelling switches, 127, 129, 138, 139, 140, 197
Levelling systems *see* Gyroscope levelling
Lever angle, 15
Lever length, 15
Lever mechanism, 10
Linear scale, 24
Liquid damping, 158
Liquid expansion compensation, 159, 164
Liquid-expansion thermometer, 267
Liquid-level switch *see* Levelling switches
Localizer, 37, 355, 357
Location of tank units, 323
Location of thermocouples, 289
Logarithmic scale, 25
Long-reach thermocouples, 289
Lubber line, 158, 160

MACH NUMBER, 91
Machmeter, 91
Magnesyn system, 248 *see also* Synchro systems
Magnet systems, 158
Magnetic compass *see* Direct-reading compasses
Magnetic dip
  definition, 156
  effect on a compass, 161
  compensation, 162
Magnetic equator, 156, 157
Magnetic field, 150
Magnetic field strength, 152
Magnetic flux, 150, 151
Magnetic foci, 156
Magnetic inclination *see* Magnetic dip
Magnetic indicators, 47
Magnetic intensity *see* Total force
Magnetic meridian, 155
Magnetic moment, 152
Magnetic poles, 150, 155
Magnetic screening, 151
Magnetic variation, 156
Magnetization curve *see* B/H curve
Magnetizing force, 182
Magnification of mechanisms, 15
Manganin spool, 294
Manifold pressure gauges, 336
Master directional gyro, 189
Master gyroscope reference, 187
Maximum safe airspeed indicator, 94
Measuring element, 10
Mechanisms
  double-tangent, 16

Mechanisms—(*contd.*)
  lever, 10
  rod, 15
  sine, 16
  skew-tangent, 17
  tangent, 16
Melting point, 266
"Mental focus" lines, 43
Mercury barometer, 70
Mercury switches *see* Levelling switches
Metering units, 101
Micro-adjuster *see* Deviation compensation devices
Micro-Desynn, 235
Micro-Desynn pressure transmitter, 305
  see also Desynn system
Microfarad, 314
Mounting of instruments, 45
Moving-coil indicator, 278
Moving-tape display, 29
Mutual inductance, 240

NEWTON PER WEBER, 152
Non-linear scale, 24 *see also* Square-law scale
Non-uniform magnetic fields, 280–283
Northerly turning error, 170
Nozzle guide vane thermocouple, 289
  see also Thermocouple types
Null point, 276
Null position, 191, 193, 194, 195

OHM'S LAW
  definition, 268
  parallel circuit, 270
  series circuit, 269
  series-parallel circuit, 272
Operating range, 27
Ozonosphere, 69

PARALLAX ERRORS, 27
Parallel circuit *see* Ohm's Law
Parasitic e.m.f., 285
Peltier effect, 285 *see also* Thermocouple principle
Pendulosity errors, 140
Pendulosity error compensation, 141
Percentage tachometer, 259
Percentage thrust indicator, 343
Period of a magnet, 154
Permalloy, 180
Permanent magnetism, 154
Permeability, 182
Permittivity, 314
Phase quadrature, 131, 245
Picofarad, 314

Pillar lights, 51
Pitch-bank erection, 139
Pitot heating circuits, 61
Pitot pressure, 57
Pitot-static pipelines, 67
Pitot-static system, 8, 55
Pitot-static tube *see* Pressure head
Pitot tube *see* Pressure head
Plasteck system, 53
Polar navigation, 200
Pole strength, 152
"Porous pot" *see* Metering units
Position error *see* Pressure error
Power follow-up, 191
Power follow-up synchro, 191, 244, 339
Power indication instruments, 335
Power loss indicator, 343
Precession, 111, 113, 115
Precession motor system, 194
Pressure, 296
Pressure altitude, 75
Pressure error, 62, 108
Pressure head, 57, 59, 60
Pressure measurement
  bellows, 301
  bourdon tube, 299
  capsules, 300
  dead-weight tester, 297
  diaphragms, 300
  methods, 296
  U-tube manometer, 296
Pressure switches, 311
Pressure transducer, 107
Pressure transmitter, 305
Pressure/weight balancing, 297
Pyrometry, 268

"Q" CODE, 78
Qualitative display, 23, 25
Quantitative display, 23

RADIATION, 266
Radio magnetic indicator, 354, 357
Radius of gyration, 113
Range markings, 33
Rapid-response thermocouple, 287 *see also* Thermocouple types
Rare-metal thermocouples, 285 *see also* Thermocouple materials
Rate of climb indicator *see* Vertical speed indicators
Rate gyroscope, 142
Ratiometer pressure gauges, 307 *see also* Desynn system and Ratiometer system

Ratiometer system, 280
Real A, 214
Recovery factor, 289
Reluctance, 151, 183
Remanance, 182
Repeater synchro, 244
Requirements for instruments, 7
Resistance bulb *see* Temperature sensing elements
Resistance thermometry, 268, 273
Resistivity, 273
Resolver synchro, 245
Rod components *see* Aircraft magnetic components
Rod mechanisms, 15
Rotation indicators, 264
Rotorace bearings, 197
R.P.M. indicators *see* Tachometers

SATURATION POINT, 182
Scale base, 23
Scale length, 25
Scale marks, 23
Scale range, 27
Scissor magnets *see* Deviation compensation devices
Sealing of instruments, 22
Sector gear, 17
Secular change, 156
Seebeck effect, 284 *see also* Thermocouple principle
Self-inductance, 240
Series circuit *see* Ohm's law
Series-parallel circuit *see* Ohm's law
Servo altimeter, 79
Servo synchro, 191 *see also* Synchro systems
Short-reach thermocouple, 289
Sine mechanism, 16
Skew-tangent mechanism, 17
Slab-Desynn, 236
Slaved gyro mode *see* Compass operating modes
Slaving amplifier, 191
Soft iron, 154 *see also* Soft iron magnetism
Soft iron magnetism, 203 *see also* Aircraft magnetic components
Space gyroscope *see* Free gyroscope
Spinning freedom, 111
Spring-rate, 301
Square-law compensation, 87
Square-law scale, 25, 87 *see also* Non-linear scale
Stagnation point, 57

Stagnation thermocouple, 287 *see also* Thermocouple materials
Standard atmosphere, 69
Standard turn rates, 143
Static counter display, 31
Static tube, 57
Static vents, 63
Steam point, 266
Step-by-step transmission system, 238
Straight scale, 29
Stratopause, 69
Stratosphere, 69
Subsonic speed, 91
Supercharging, 336
Supersonic speed, 91
Surface-contact thermocouple, 285 *see also* Thermocouple types
Synchro systems
  basic, 241
  definition, 239
  Magnesyn, 248
  power follow-up synchro, 191, 244
  repeater synchro, 244
  resolver synchro, 245
  servo synchro, 191
  synchrotel, 247
Synchronizing indicators, 199
Synchroscope, 35, 260
Synchrotel, 247

TACHOMETERS
  centrifugal, 252
  drag elements, 257, 259
  electrical, 253
  flexible drive shafts, 252
  functions, 252, 336
  generators, 255
  indicators, 256
  percentage tachometer, 259
  temperature compensation, 259
Tangent mechanism, 16
Tank units
  characterized, 324
  compensator, 323
  float-type, 312
  standard (capacitance), 323
Temperature and scales, 266
Temperature bulb *see* Temperature sensing elements
Temperature coefficient, 273
Temperature compensation methods
  bimetal strip, 19, 290
  thermo-magnetic shunt, 21, 259, 293
  thermo-resistance, 21
Temperature control system, 347

Temperature effects on fuel, 319
Temperature scale conversions, 267
Temperature measurement, 266
Temperature/resistance laws, 273
Temperature and r.p.m. control system, 349
Temperature sensing elements, 276
Terrestial magnetism, 155–158
Thermistor *see* Thermo-resistor
Thermocouple
  combinations, 285
  location, 289
  materials, 285
  principle, 284
  recovery factor, 289
  types, 285–289
Thermomagnetic shunt, 21, 259, 293
Thermometer display *see* Moving-tape display
Thermo-resistor, 21, 293
Thomson effect *see also* Thermocouple principle, 285
"Tied" gyroscope *see* Earth gyroscope
Tilting freedom, 111
"Top limiting", 351
Toppled gyro, 133 *see also* "Tumbling" of a gyroscope
Toroidal resistance, 230
Torquemeter, 338
Torque motors, 127, 129, 195, 197
Torque motor erection system, 124
Torque pressure indicators, 338, 339
Torricellian vacuum, 71
Total force, 157
Total pressure *see* Pitot pressure
"Trace" disc, 73
Transformer principle, 240
Transonic range, 91
Trimming resistance, 294
Tropopause, 69
Troposphere, 69
True altitude, 77
True poles *see* Geographic poles
"Tumbling" of a gyroscope, 122 *see also* Toppled gyro
Tuning spring, 89
Turbine temperature control, 345
Turn-and-bank indicators, 142–148
Turning errors
  compasses, 168
  gyro horizons, 135, 140
Turns ratio, 241

ULTRA-VIOLET FLOOD LIGHTING, 50
Units of capacitance, 314

U-tube manometer, 296

VACUUM-DRIVEN GYRO HORIZON, 122
Vapour-pressure thermometer, 267
Veeder-counter display *see* Digital display
Veering freedom, 111
Verge glass, 165
Verge ring, 165
Vertical gyroscope, 119
Vertical speed indicators, 97
Vibration monitoring, 351

Vibration pick-up, 351
Viscosity compensator valve, 102
V.O.R., 354, 357, 359

WATER TRAPS *see* Drains
Weber, 150
Wedge-type lighting, 52 *see also* Instrument illumination
Wheatstone bridge network, 275

ZERO ADJUSTMENTS, 19, 102